CHAPTER MA

(1901)

Being the Opening, Closing, Secret Work and Lectures of the Mark Master, Past Master, Most Excellent Master and Royal Arch Degrees. Illustrated.

Edmond Ronayne

ISBN 1-56459-746-6

Kessinger Publishing's Rare Reprints
Thousands of Scarce and Hard-to-Find Books!

PREFACE.

Occupying successively the official positions of Secretary, Senior Warden and Worshipful Master of Keystone Lodge No. 639, Chicago, it became imperative that I acquire a thoroughly accurate knowledge of the Standard ritual and work of the three symbolic degrees; but that was easily accomplished by witnessing how the degrees were conferred in other lodges, by occasional attendance at the sessions of the Grand Lecturers; by observing the work as exemplified in the Grand Lodge (of which I was a member) and above all by personal instruction received from such prominent Masons as D. H. Kilmore, H. F. Holcomb, and John O'Neil, District Deputy Grand Masters in Chicago, but more especially from Mr. Edward Cook, late Grand Master of Masons of Illinois.

Having taken the Chapter degrees some years before affiliating with Keystone, but paying only slight attention to that particular branch of Masonry, I also determined to become equally proficient in the "secret work" of the Chapter as in that of the Blue Lodge. With that intent I gladly availed myself of the proposed assistance of personal friends among leading Royal Arch Masons, with whom I used to rehearse as time would permit the various parts of the ritual and lectures, until the whole were thoroughly understood and memorized. Ambitious to be looked up to as a "bright Mason," was my chief incentive in those days, but the Masonic knowledge then acquired has been of the greatest benefit in my public lectures, while constant study, a perfect familiarity with Masonic phraseology, and above all a most retentive memory, enable me now to write out the entire "secret work" of Chapter Masonry with the same accuracy of detail as that recently manifested, when with the assistance of twenty-seven other seceding Masons I publicly worked the first and third degrees of the Blue Lodge, here in Chicago, in the presence of four thousand people.

I am gratefully indebted, however, to the personal friendship of a prominent member of York Chapter for those important amendments to the ritual lately authorized by the General Grand Royal Arch Chapter—now published for the first time—and which the intelligent Companion Royal Arch Mason cannot fail to recognize and appreciate.

My "Hand Book of Freemasonry," written immediately upon leaving Keystone Lodge, and containing the Standard ritual and work of the first three degrees, has not only furnished the public at large with an accurate revelation of the alleged secrets and mysteries, together with the oaths and death penalties of the Masonic system, but it has also been in general private use from the very first, as an absolutely reliable Text Book among the members of the Craft throughout the country.

For years there has been an increasing demand for an equally reliable exposition of the four Chapter degrees, and to supply that demand as well as in response to numerous inquiries from time to time, the present work has been written. It is also intended as a companion to the "Hand Book" It is completely illustrated. every diagram and figure having been expressly prepared for these pages. In preparing the present work, my great object has been, to give Chapter Masonry with such absolute accuracy as not only to challenge the closest criticism of the best informed Royal Arch Masons, but also to produce a work upon the correctness of which the public at large can perfectly rely.

Under these circumstances, then, the present volume is sent forth in the Name of Him whom the Masonic system rejects, praying that the spiritual hoodwinks may be removed, the vails of darkness, delusion and avarice drawn aside, and God's people, if any there be in the lodge and Chapter, speedily released from the unequal yoke of the cable-tow, while the mere American citizen, apart altogether from any Christian obligation, may be led to realize the folly, illegality and crime involved in the awful oaths and horrible death penalties of the Masonic order, and so become a *free man* by severing his connection forever with what even thousands of Masons throughout the land privately admit to be one of the most gigantic and selfish *trusts* of modern times.

E. RONAYNE.

104 Milton Avenue,
    Chicago, Ill.

ENDORSEMENTS.

I have carefully examined the Proof Sheets of "Chapter Masonry" by E. Ronayne giving the ritual secret work and lectures of the four degrees of the Chapter. namely "Mark Master," "Past Master," "Most Excellent Master" and "Royal Arch," and have much pleasure in testifying to the absolute correctness of the same to the best of my knowledge and recollection.

B. F. STANTON,

Ex-Captain of Host,

Vandalia Chapter,

Vandalia, Ills.

I have examined the Proof Sheets above referred to and cheerfully endorse what Ex-Companion Stanton has said.

SILAS C. BURNETT,

Ex-Member of Excelsior Chapter No. 40 Royal Arch Masons,

Larned, Kans.

ILLUSTRATIONS.

CHAPTER I.

CHAPTER MASONRY.

DEGREE OF MARK MASTER MASON.

THE term Chapter when used Masonically is, strictly speaking, the name applied to a series of degrees in the system of Freemasonry, just as Blue Lodge, Council, and Commandery, are the names of ther series in the same connection. ·

In the Blue Lodge series there are three degrees: The Entered Apprentice, Fellow Craft and · Master Mason, and these in reality constitute the only true and legitimate Masonry there is. The ceremonies in these degrees were *Revived* from the old initiatory systems of the ancient pagan mysteries, while all the other degrees, no matter to what series or systems they belong, were *manufactured* from about the middle to the end of the 18th. Century. (See "Master's Carpet," pp. 199-210.) ,

The Chapter has four degrees: Mark Master, Past Master, Most Excellent Master, and Royal Arch, and these will come before us now more particularly for illustration and comment. The *secret work* of each degree as set forth by the General Grand Chapter of the United States will be given, including the proper manner of opening, closing and conferring the degrees, together with the *lecture* pertaining to each; the whole furnishing a ready and reliable means, not alone to Masons, but also to all others, of becoming practically familiar with the mischief and mysteries of the Chapter degrees.

In Blue Lodge (or Symbolic) Masonry all routine business must be transacted in the Master's degree; lodges in the other two degrees being opened for work only. So, likewise, in the Chapter all business must be done in the highest, or Royal Arch degree; the lodges below that—the Mark, Past, and Most Excellent Master—being always opened for initiatory purposes alone, these degrees being conferred under the authority of the Royal Arch, which in that connection is termed the Chapter by way of eminence; and hence all four degrees in the series are termed Chapter degrees.

The officers of the Chapter under whose warrant, or charter, the Mark Master's degree is conferred, are as follows:

Excellent High Priest, King and Scribe; Captain of the Host; Principal Sojourner; Royal Arch Captain; Master of the 3rd. Vail; Master of the 2nd. Vail; Master of the 1st. Vail; also a Treasurer and Secretary, as in other lodges; and a Sentinel, or Tyler.

In the Mark Master's degree these officers rank as follows:

Excellent High Priest as Right Worshipful Master.

Excellent King   as Senior Warden.
Excellent Scribe   as Junior Warden.
Principal Sojourner   as Senior Deacon.
Royal Arch Captain   as Junior Deacon.
Master of the 3rd. Vail as Master Overseer.

Master of the 2nd. Vail as Senior Overseer.

Master of the 1st. Vail as Junior Overseer.

Captain of the Host as Marshal, or Master of Ceremonies.

Master Masons only are eligible for the degree of Mark Master, and the candidate must be balloted for in the Chapter at a regular convocation. If elected to receive the degree, he is notified when to appear for initiation, but before the degree is conferred he must be examined in the *Lecture* of the Master Mason's degree, as found in the "Hand Book of Freemasonry," pages 258 to 269, inclusive; and hence it is absolutely necessary that he make himself familiar with that lecture.

The symbolic *color* of the first three degrees is *blue;* hence the term Blue Lodge Masonry. The characteristic color of the Chapter is *scarlet*, and the regalia and jewels of the Royal Arch are those which are used in the three preceding degrees.

The meeting of the Blue Lodge is termed a "communication;" that of the Chapter a "convocation;" and in either case are said to be "regular" or "emergent." At a regular meeting both routine business and initiations can be conducted; while at an emergent meeting only that particular business for which it has been called can be transacted.

In order that the general reader, as well as the intelligent Mason, may have a correct understanding of how the Chapter degrees are conferred, I shall take each degree in its order, first *opening* the lodge or chapter "in due and ancient form;" then *giving the secret work* of the degree; next, the lecture, and, finally, closing the lodge. Thus every degree will comprise four chapters—opening, working, lecturing and closing; and in addition to that it will be shown how lodges are raised or lowered from one degree to another as occasion may require. ·

The Master's degree is based upon the legend of Hiram Abiff; the Mark Master's degree on the legend of a Keystone. Both legends are absolutely false, the various statements being both improbable and impossible, as any intelligent Mason may know by reading the account of how all material for King Solomon's Temple was provided. (1 Kings, Chapters VI and VII.)

With these preliminary remarks I shall now proceed to open a lodge of MARK MASTER MASONS.

OPENING CEREMONIES.

The brethren having assumed their appropriate regalia, the Right Worshipful Master gives one rap with his gavel; the officers take their respective stations, and brethren are seated as in the accompanying diagram:

("Royal Arch Standard," 1897, p. 12.)

R. W. Master: "Brother Senior Warden, are you satisfied that all present are Mark Master Masons?"

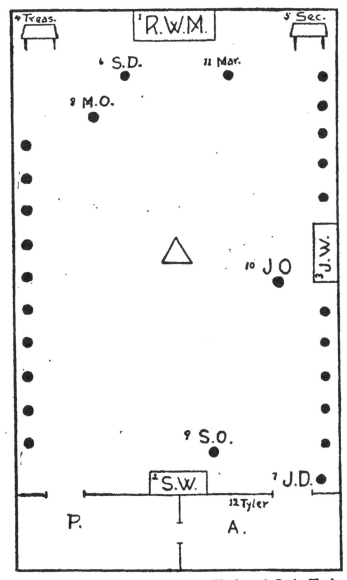

1. Right Worshipful Master. 2. Senior Warden. 3. Junior Warden.
4. Treasurer. 5. Secretary. 6. Senior Deacon. 7. Junior Deacon. 8. Master Overseer. 9. Senior Overseer. 10. Junior Overseer. 11. Marshal.
12. Tyler or Sentinel.

S. Warden (rising): "I will ascertain by my proper officers and report. Brothers Senior and Junior Deacons, approach the west."

These officers meet at the west of the Altar, and repair to the Senior Warden's station.

"Give me the pass of a Mark Master Mason."

They approach the Senior Warden's station and whisper in his ear the word Joppa.

S. W : "You will now ascertain if all present are Mark Master Masons and report to the west."

The deacons then pass along in front of the brethren on the north and south sides of the lodge and each brother, as the deacon approaches, *rises* and whispers in his ear the *password* of this degree, which is *Joppa.* (See p. 51) Having thus collected the pass from all but the Master, Wardens and Overseers, they return to the west and report to the Senior Warden.

Senior Deacon: "Bro. Senior Warden, all present are Mark Masters in the north."

Junior Deacon: "Bro. Senior Warden, all present are Mark Masters in the south."

The Senior Warden then reports as follows:

S. W.: "R. W. Master, all present are Mark Master Masons;" and takes his seat.

This ceremony which is Masonically termed, "purging the lodge," must never be omitted in opening any lodge of Masons, and after it is performed, the due guard and sign of the degree on which the lodge is at work are always given, on addressing the Right Worshipful Master.

R. W. M. (one rap): "Brother Junior Deacon, what is the first great care of Masons when in lodge assembled?"

Junior Deacon (rising): "To see that the lodge is duly tyled, Right Worshipful Master."

R. W. M.: "Perform that duty and inform the Tyler that we are about to open a lodge of .Mark Master Masons; direct him to tyle accordingly."

The Junior Deacon opening the ante-room door ajar, tells the Tyler in a low voice that a Mark Master's lodge is to be opened, closes the door and gives two knocks in quick succession; then pauses a little and gives two other knocks in the same way.

These knocks are answered in a similar manner by the Tyler; after which the Junior Deacon reports:

Junior Deacon: "The lodge is duly tyled, Right Worshipful."

R. W. M.: "How is it tyled?"

J. D.: "By a worthy brother without, armed with the proper implement of his office."

R. W. M.: "What are his duties?"

J. D.: "To keep off all cowans and eaves-droppers, and admit none but such as are duly qualified, and have permission from the Right Worshipful Master."—The Junior Deacon then takes his seat.

R. W. M. (one rap): "Bro. Senior Warden, are you a Mark Master Mason?"

S. Warden (rising): "I am; try me."

R. W. M.: "How will you be tried?"

S. W.: "By the chisel and mallet."

R. W. M.: "Why by the chisel and mallet?"

S. W.: "Because they are the working tools of a Mark Master Mason."

R. W. M.: "Where were you made a Mark Master Mason?"

S W.: "In a legally constituted and duly opened lodge of Mark Master Masons."

R. W. M.: "How many compose such a lodge?"

S. W.: "Eight or more."

R. W. M.: "When composed of eight, who are they?"

S. W.: "The Right Worshipful Master, Senior Warden, Junior Warden, Senior Deacon, Junior Deacon, Master, Senior and Junior Overseers."

R. W. M.: "The Junior Overseer's place in the lodge?"

S W.: "At the South gate." (Senior Warden is seated.)

R. W. M. (One rap. Junior Overseer rises and gives the due guard): "Your duty there, brother Junior Overseer?"

Junior Overseer: "To inspect work presented at the South gate."

R. W. M.: "The Senior Overseer's place?"

J. O.: "At the West gate, Right Worshipful."

R. W. M. (One rap. The Senior Overseer rises and gives the due guard): "Your duty there, brother Senior Overseer?"

Senior Overseer: "To inspect work presented at the West gate."

R. W. M.: "The Master Overseer's place, brother Senior Overseer?"

S. O.: "At the East gate, R. W. Master."

R. W. M. (One rap. Master Overseer rises and gives the due guard): "Your duty there, brother Master Overseer?"

Master Overseer: "To complete the inspection of work when presented at the East gate, Right Worshipful."

R. W. M. "Brother Master Overseer, the Junior Deacon's place?"

Master Overseer: "At the right of the Senior Warden, in the West."

R. W. M. (One rap. Junior Deacon rises and gives the due guard): "Your duty there, brother Junior Deacon?"

Junior Deacon: "To carry messages from the Senior Warden in the West to the Junior Warden in the South and elsewhere about the lodge as he may direct, and to see that the lodge is duly tyled."

R. W. M.: "The Senior Deacon's place?"

J. D.: "At the right of the Right Worshipful Master "

R. W. M. (One rap. Senior Deacon rises and gives the due guard): "Your duty there, brother Senior Deacon?"

Senior Deacon: "To carry orders from the Right Worshipful Master in the East to the Senior Warden in the West and elsewhere about the lodge, as he may direct; to welcome and accommodate visiting brethren, and to receive and conduct candidates."

R. W. M.: "The Junior Warden's station?"

S. D.: "In the South, Right Worshipful."

R. W. M. (Two raps. All the officers rise, and the Junior Warden gives the due guard): "Why in the South, and your duty there, brother Junior Warden?"

Junior Warden: "As the sun in the South at meridian height is the glory and beauty of the day, so is the Junior Warden in the South, the better to observe the time, to call the craft from labor to refreshment, superintend them during the hours thereof, see that none convert the purposes of refreshment into intemperance or excess, and call them on at the will and pleasure of the Right Worshipful Master."

R. W. M.: "Brother Junior Warden, the Senior Warden's station?"

J. W.: "In the West, Right Worshipful."

R. W. M.: "Why in the West, brother Senior Warden, and your duty there?"

Senior Warden: "As the sun is in the West at the close of the day, so is the Senior Warden in the West to assist the Right Worshipful Master in opening and closing the lodge, to pay the craft their wages, if aught be due, and see that none go away dissatisfied; harmony being, the support of all well-governed institutions."

R. W. M.: "The Right Worshipful Master's station?"

S. W.: "In the East, Right Worshipful."

R. W. M.: "Why in the East, and his duties there?"

S. W.: "As the sun rises in the East to open and govern the day, so is the Right Worshipful Master in the East to open and govern the lodge, set the craft to work and give them proper instruction for their labor."

R. W. M. (Gives three raps. All the brethren stand): "Brother Senior Warden, it is my order that a lodge of Mark Master Masons be now opened for the dispatch of business under the usual Masonic restrictions. This order you will communicate to the Junior Warden in the South, and he to the brethren for their government."

S. W.: "Brother Junior Warden, it is the order of the Right Worshipful Master that a lodge of Mark Master Masons be now opened for the dispatch of business under the usual Masonic restrictions. This order you will communicate to the brethren present for their government."

Junior Warden: "It is the order of the Right Worshipful Master that a lodge of Mark Master Masons be now opened for the dispatch of business under the usual Masonic restrictions; of this you will take due notice and govern yourselves accordingly. Look to the East."

The brethren now, following the lead of the R. W. M., make *all* the signs, from the Entered Apprentice degree to that of Mark Master Mason, inclusive, as represented in accompanying figures.*

R. W. M. gives two quick raps; Senior Warden two, Junior Warden two.

R. W. M. again two raps; Senior Warden two, and Junior Warden two.

* See following pages.

FIRST DEGREE.   ## SECOND DEGREE.

Due Guard.   Penal Sign.   · Due Guard.   Penal Sign.

THIRD DEGREE.   ## DEGREE OF MARK MASTER.

Due Guard.   Penal Sign.   Heave Over.   Grand Hailing Sign.

After which the Right Worshipful Master, removing his hat, reads the following mutilated portion of Scripture; or a minister, if present, may read it; and from which, by Masonic law, the name of Jesus Christ is knowingly and wickedly excluded.

DEGREE OF MARK MASTER.

Due Guard.   Penal Sign.

"CHARGE TO BE READ AT OPENING THE LODGE."

"Wherefore, brethren, lay aside all malice, and guile, and hypocrisies, and envies, and all evil speakings.

"If so be ye have tasted that the Lord is gracious; to whom coming, as unto a living stone, disallowed indeed of men, but chosen of God, and precious; ye also as living stones be ye built up a spiritual house, an holy priesthood, to offer up sacrifices acceptable to God.

"Wherefore, also, it is contained in the Scriptures, Behold I lay in Zion, for a foundation, a tried stone, a precious corner-stone, a sure foundation; he that believeth shall not make haste to pass it over. Unto you, therefore, which believe, it is an honor; and even to them which be disobedient, the stone which the builders disallowed, the same is made the head of the corner.

"Brethren, this is the will of God, that, with well-doing, ye put to silence the ignorance of foolish men. As free, and not using your liberty for a cloak of maliciousness, but as the servants of God. Honor all men; love the brotherhood; fear God."—1 Peter 2: 1, 7, 15, 17.*

R. W. M. (Continuing): "I now declare this lodge of Mark Master Masons, erected to God and dedicated to the memory of Hiram Abiff, opened in due form. Brother Junior Deacon, inform the Tyler. Brother Senior Deacon, display the three great lights."

Junior Deacon gives four raps on Ante-room door, each two raps in quick succession, thus: **—** These are answered by the Tyler in the same manner. Junior Deacon opens the door, and announces in a low voice to the Tyler that the lodge is open. He closes the door, and again raps as before, which the Tyler answers as before; the

J. D. facing the East and saluting with the due guard and sign: "The Tyler is informed, Right Worshipful." In the mean time the Senior Deacon places the square and compass on the open Bible,

---

* Royal Arch Standard, 1897, p. 13.

and facing the Right Worshipful Master, also salutes with the due guard and sign.*

R. W. M. gives one rap, and the brethren are seated as on page 5.

There are four regular signs in all, connected with the Mark Master's degree, and are given in their order as follows: The Heave Over, the Grand Hailing Sign, the Due Guard, and the Penal Sign.

1. The Heave Over alludes to the rejection of the Keystone, and is made by placing the open palm of the right hand in that of the left hand near the right hip, both palms open and turned upward. Then raise them with a quick motion, as if heaving something over the left shoulder. See Figure p. 12.

2. The Grand Hailing Sign is made by extending right arm with the two first fingers and thumb of the hand outstretched, the other two fingers clinched. It alludes to the manner of carrying the Keystone between the thumb and two first fingers, and is also the manner of receiving wages. See Figure p. 12.

3. The Due Guard is made by passing the two first fingers of the right hand over the right ear as if pushing back the hair, and alludes to a part of the penalty of the obligation, to have the right ear cut off. See Figure p. 13.

4. The Penal Sign is made by extending the right arm, palm upward, bringing lower edge of left hand open palm down on the right wrist, as if chopping off the right hand. It alludes to the additional part of the penalty—to have the right hand chopped off. See Figure p. 13.

* The due guard ought to be given by a brother or an officer when addressing the Right Worshipful Master.

CHAPTER II.

MARK MASTER'S DEGREE.

THE SECRET WORK.

The non-Masonic reader will need to be informed that in each degree of Masonry proper the initiatory ceremonies are divided into two or more sections, aside from the lecture or catechism pertaining to each section. By referring to the "Hand Book of Freemasonry" this fact will become apparent as regards the degrees of the Blue Lodge. The Chapter degrees are likewise divided into sections, each of which shall be specially referred to as we proceed. The degree before us—that of Mark Master—is divided into two sections, and we shall now proceed to illustrate the revised ritual of the

FIRST SECTION.

The lodge being opened as in the preceding chapter, the degree work proceeds as follows:

Right Worshipful Master (rising): "Brethren, this lodge of Mark Master Masons has been called and opened for the purpose of advancing to the honorary degree of Mark Master, Brother James

Hunt, who has been duly elected to receive the same, and if there be no objection we will now proceed to confer the degree." No objection being made, he continues: "There being none, it is so ordered."

R. W. M.: "Brother Senior Deacon, you will ascertain if the candidate is in waiting."

The Senior Deacon goes to the west side of the altar and salutes the Right Worshipful Master with the due guard and sign. See p. 13. He then repairs to the ante-room door, upon which the Junior Deacon gives four raps by two's. (** **), which are answered in the same manner by the Tyler. The J. D. opens the door, and the S. D. passes out. On returning, the Tyler knocks first, and is answered by the J. D. The Tyler opens the door, and the S. D., entering the lodge, proceeds to the west side of the altar, salutes as before with due guard and sign, (p. 13,) and reports:

S. D : "Right Worshipful Master, I find in waiting Brother James Hunt, who has been duly elected to take the Mark Master's degree." He then resumes his seat.

It may be well to explain once for all that whenever any one is retiring from the lodge the Junior Deacon knocks upon the ante-room door and opens it, but when anyone is entering the Tyler knocks and opens the door.

R. W. M.: "Brother Secretary, has the candidate paid the required fee?"

If the fee has been paid (or remitted when the candidate is a preacher,) the Secretary replies:

Secretary: "He has, Right Worshipful."

But if not paid, or remitted, the Secretary is ordered by the R. W. M. to retire and collect it, which he does—saluting as usual, with due guard and sign, both in retiring from and entering the lodge, and reports:

Secretary: "The usual fee has been collected, Right Worshipful."

R. W. M.: "Brother Senior and Junior Deacons, you will retire, prepare and introduce the candidate."

The two Deacons go at once to the west side of the altar, and facing the East, salute together the Right Worshipful Master with the due guard and sign of a Mark Master, and then proceed to the preparation room.*

PREPARATION OF CANDIDATE.

The candidate is deprived of all money, coin or paper; his coat is removed; both sleeves are rolled up above the elbows, and an apron is put on him, which he wears as a Fellow Craft Mason. The Senior and Junior Deacons are similarly prepared, and each of these officers carries in his right hand an oblong stone—or what resembles a stone—about eight or ten inches long, four or five inches wide, and about an inch thick, having the representation of a square on that carried by the Senior Deacon, and a plumb on the one borne by the Junior Deacon. These stones they grasp at the top in the natural way be-

---

* Every officer and member retiring from the lodge after it is opened, must salute the Master with the due guard and sign of the degree on which the lodge is working, and the same when an officer or member enters.

tween the thumb and fingers, swinging them by their side at full arm's length.

The candidate is given a Keystone, much larger and heavier than that carried by the Deacons, on one side of which are two concentric circles, with the letters H. T. W. S. S. T. K. S. This Keystone he is directed by the Junior Deacon to take in his right hand in the ordinary way of carrying a stone—between the thumb and four fingers—holding it by the small end and swinging it at arm's length as the two Deacons do theirs, the lettered side being outward. See cuts below.

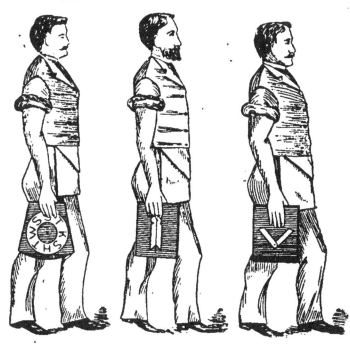

Candidate.   Junior Deacon.   Senior Deacon.

All being ready, they enter the lodge room in single file, without knocking, the Senior Deacon first, the Junior Deacon next, and the candidate last, as represented in the accompanying figures, and proceed around the lodge to the stations of the Overseers, as shown in the diagram on page 21, making the first stop at the South gate, in front of the Junior Overseer, where the Senior Deacon gives four knocks in the manner already explained---** **.

Junior Overseer. (rising): "Who comes here?"

Senior Deacon: "Craftsmen from the quarries, with work for inspection."

J. O.: "Present your work."

The Senior Deacon swings his right hand, in which he carries the stone, once from the rear to front, throwing it up level with his breast, the palm of the hand being thus turned upward and the arm extended. With his left hand he then catches the right hand and stone, the left under the right; and, of course, both palms are turned upward; the stone is thus presented for inspection.

The Junior Overseer takes the stone and applying his trying square to its corners and sides, says:

J. O.: "This is good work, true work, square work; pass to the Senior Overseer."

The Senior Deacon moves on about two or three steps and the Junior Deacon approaches the Junior Overseer.

J. O.: "Present your work."

The J. D. presents his oblong stone just as the S. D. did his, and being examined in the same way, J. O. says:

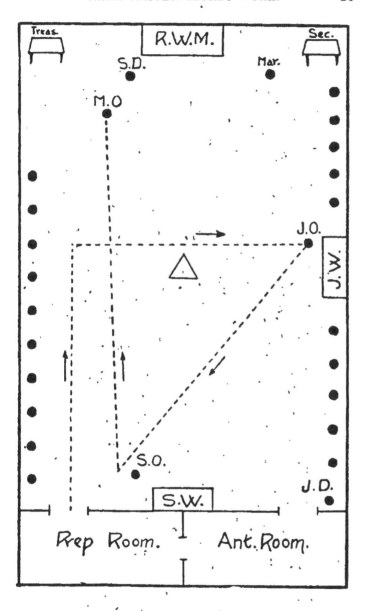

J. O.: "This is good work, true work, square work; pass to the Senior Overseer."

Senior and Junior Deacons advance toward the west a few steps and the candidate approaches the Junior Overseer, who says:

J. O.:. "Present your work."

The candidate swings his right'arm as he saw the Deacons do, bringing up the Keystone on a level with his breast, his left hand under his right, both palms upward, the large end of the Keystone being also uppermost, and the letters facing the Junior Overseer in their natural order, as illustrated in the accompanying figure. The Junior Overseer taking

the Keystone applies to it his little trying square, and finding it not to fit its sides and angles, he exclaims:

J. O.: "This is neither oblong nor square; neither has it the mark of any of the craft upon it. but, owing to its singular form and beauty, I am unwilling to reject it, and will suffer it to pass to the Senior Overseer for further inspection."

The three now move on, in the same order as before, to the West gate; where sits the Senior Overseer, and where precisely the same ceremonies occur, and the same remarks are made as at the South gate. The Senior Overseer directs them to the Master Overseer, at the East gate, for final inspection.

Arriving before the Master Overseer, the Senior Deacon gives four knocks—two and two.

Master Overseer: "Who comes here?"

S. D.: "Craftsmen from the quarries, with work for inspection."

M. O.: "Present your work."

The Senior Deacon presents his work in the manner already explained, and the Master Overseer, applying his square as the others did, exclaims:

M. O.: "This is good work, true work, square work, and has a proper mark upon it, and entitles you to wages."

Senior Deacon moves forward a short distance and the Junior Deacon approaches.

M. O.: "Present your work."   (J. D. presents it, as before.)

M. O.: "This is good work, true work, square work, and has a proper mark upon it, and entitles you to wages."

The Junior Deacon moves on a short distance, and the candidate approaches the Master Overseer.

M. O.: "Present your work."

Candidate presents the Keystone as already explained, and the Master Overseer taking it, applies his trying square to its sides and angles, which of course does not fit them; he then pretends to examine it critically, and with apparent surprise, and says:

M. O.: "This is neither oblong nor square, neither has it the mark of any of the Craft upon it. Stand aside. Overseers assemble."

Junior and Senior Overseers repair to the East gate, as the Master Overseer's place is termed.

M. O.: "Brother Junior Overseer, did you permit this piece of work to pass your inspection?"

J. O.: "I did, with the remark at the time that it was neither oblong nor square, but owing to its singular form and beauty I was unwilling to reject it, and permitted it to pass to the Senior Overseer for further inspection."

S. O.: "And I, for similar reasons, permitted it to pass on to you for final inspection."

M. O.: "Brethren, you see it is neither oblong nor square; square work only is such as we have orders to receive; neither has it the mark of any of the Craft upon it known to me. Do you know that mark?" (pointing to the circles and letters on the Keystone,) "or any use for such a stone?"

S. O.: "I do not."

J. O.: "Neither do I."

M. O.: "Neither do I. What shall we do with
t?"

. J. O.: "Let us heave it over among the rubbish."

M. and S. O. (together): "Agreed."*

The Master and Senior Overseers now take the
Keystone between them, and swinging it four times
back and forth quite slowly, the fourth time they
leave it over the left shoulder of the Senior Overseer, where the Junior Overseer receives it and
deposits it in any convenient place for future use—
usually near the Senior Warden's pedestal. Heav-
ng the Keystone over the left shoulder furnishes the
leave over sign. (See p. 12.)

M. O. (to candidate): "Your work is rejected,
and you are not entitled to wages."

Senior Deacon leads candidate to a seat near
the North East corner of the lodge, and both Deacons then repair to their respective places.

R. W. M. (one rap): "Brother Junior Warden
what is the hour?"

J. W. (rising): "This is the sixth hour of the
sixth day of the week, Right Worshipful."

R. W. M.: "This is the day and the hour when
the Craftsmen should repair to the apartments of the
Senior Warden to receive their wages. You will
give your orders accordingly."

* They were nice men to be placed as overseers of work for the
Temple, and by the wise Solomon, too! Evidently they knew nothing of
the arch in architecture, even if used at that time.

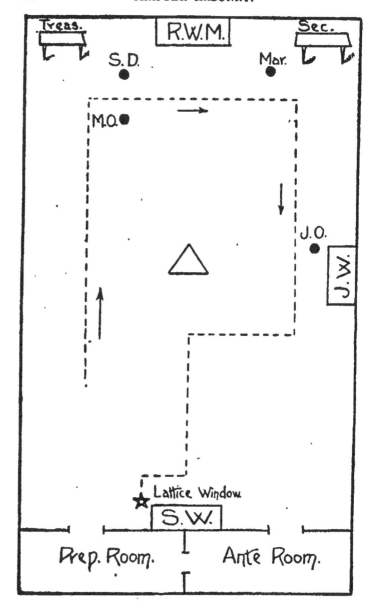

J. W. (three raps and all the brethren rise): "Craftsmen, you will assemble and repair to the apartments of the Senior Warden to receive your wages."

Master of Ceremonies: "Brethren form in procession on the North side of the lodge, single file, facing the East."

The procession is formed as directed by the Master of Ceremonies or Marshal, the Master, Senior and Junior Overseers being in front; after them, are the Craftsmen or brethren, then the Senior Deacon, and last of all the candidate. In the mean time a "Lattice Window" has been attached to the dais or platform on which the Senior Warden sits, and there the Junior Deacon is stationed with a drawn sword raised above the lattice work.

The procession moves East, South, and towards the West, as in the diagram on p. 26; each Craftsman as he arrives at the Senior Warden's station thrusting his right hand through the opening in the Lattice Work as if to receive his wages. When the candidate comes up he does the same, but not being yet instructed as to the proper manner of receiving wages, his right hand is seized by the Senior Warden and pulled further in. (See figure, page 28.)

S. W. (holding candidate's hand): "An imposter! an imposter! Strike off his right hand!"

The Junior Deacon brings down the sword as if to obey this order, as illustrated in accompanying figure of the Lattice Window.

S. D. (standing near the candidate): "Hold! he is no imposter."

S. W.: "He has attempted to draw wages when none were due him."

S. D.: "I know him to be a Fellow Craft. I have wrought with him in the quarries."

S. W.: "Can you vouch for him as a Fellow Craft?"

S. D.: "I can."

S. W.: "I will then release him, upon the condition that you will take him to the Right Worshipful Master for his decision."

S. D.: "I will receive him upon that condition."

The candidate is released by the Senior Warden, and is conducted by the Senior Deacon to the Right Worshipful Master in the East. All other officers and brethren take their seats.

S. D.: "Right Worshipful Master, this young Craftsman has just been detected as an imposter at the apartments of the Senior Warden, in attempting to draw wages when none were his due; but he was released upon condition that I present him before you for examination and decision."

M. O.: "Right Worshipful Master, he has also been guilty of presenting work that would not pass inspection."

R. W. M. (addressing candidate): "Are you a Fellow Craft?"

Candidate: "I am; try me."

R. W. M.: "Give me the due guard and sign of a Fellow Craft."

The candidate gives the due guard and sign of a Fellow Craft, as in the accompanying figures. ("Hand Book of Freemasonry," p. 33, 34.)

Due Guard.   Sign.

R. W. M.: "You appear to be a Fellow Craft. Can it be possible that it was your intention to impose upon the Craft? Do you know the penalty of an imposter?"

Candidate: "I do not."

R. W. M.: "That serves in some measure to mitigate the offence, and I will remit the penalty upon the condition that on the first hour of the first day of the week, when the Craftsmen return to the quarries to resume their labors, you go with them and there labor until you shall be able to present such work as will pass the Overseers' squares and will entitle you to wages.

"My brother, these ceremonies are intended for the purpose of impressing upon your mind in the strongest possible manner that a man should never under any circumstances attempt to receive that which is not his just due, nor in any manner attempt to impose upon any one, more especially a brother Mason; that should he do so he is not only guilty of a great wrong, but also violates his solemn obligations. Be careful, my brother, that thou receivest no wages, here or elsewhere, that are not thy just due, for if thou doest, thou wrongest some one by taking that which in God's chancery belongeth to him; and whether that which thou takest be wealth, or rank, or influence, or reputation."

Senior Deacon now conducts the candidate to a seat—usually, at this time, on the south side of the lodge, towards the Junior Deacon's place.

R. W. M.: (gives three raps—all rise): "Brethren, the sixth day of the week has drawn to a close and the holy Sabbath has begun; let us attend reverently to the worship of the God of our fathers."*

The Chaplain, or in his absence the Right Worshipful Master, approaches the altar and reads from the Chapter Manual:

"In six days God created the heavens and the earth, and rested upon the seventh day; the seventh, therefore, our ancient brethren consecrated as a day of rest from their labors, thereby enjoying frequent opportunities to contemplate the glorious works of the creation and adore their great Creator."

The following hymn is also sung by all the brethren to the tune of

OLD HUNDRED.

Another six days' work is done,
Another Sabbath is begun;
Return, my soul! enjoy thy rest,
Improve the day thy God hath blessed.

In holy duties let the day
In holy pleasures pass away;
How sweet a Sabbath thus to spend,
In hope of one which ne'er shall end.

R. W. M. (one rap): "Brother Junior Warden, what is the hour?"

J. W. (rising): "It is the first hour of the first day of the week."

R. W. M.: "This is the day and the hour when the Craftsmen should return to the quarries to resume their labors. You will give your orders accordingly."

---

* The God of our fathers! A gang of infidels who don't believe one word of what God has said, thus making a mockery of Scripture and turning the truth of God into a lie. The god of Masonry is Mah-hah-bone—a substitute—and not the God of our Lord Jesus Christ.

J. W. (three raps—all rise): "Craftsmen, you will return to the quarries and resume your labors."

The Junior Deacon conducts the candidate to the preparation room where he instructs him how to bring up his work for inspection. Instead of a keystone he is now handed one of the oblong stones, that formerly carried by the Senior Deacon. This he is taught to take between his two first fingers and thumb—the two last fingers being closed—the Junior Deacon takes his oblong stone in the same way. (See figures below.)

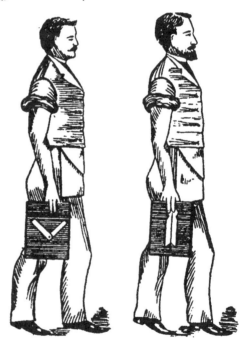

The candidate being thus instructed, both re-enter the lodge room without ceremony, and proceed at once to the Junior Overseer, to whom they present their work as before—the Junior Deacon first and the candidate next.

The Junior Overseer on examination pronounces it "good work, true work, square work," and directs them to the Senior Overseer, and he to the Master Overseer for final examination.

The Master Overseer in like manner declares it to be "good work, true work, square work," and further tells the candidate that he is "entitled to wages."

The candidate's work having thus passed the inspection of the Overseers, the Senior Deacon now takes charge of him, and approaching the East presents him to the Right Worshipful Master, thus:

S. D.: "Right Worshipful Master this is the young Fellow Craft who was ordered to the quarries, there to labor until he should be able to present such work as would pass the Overseers' squares. He has done so."

R. W. M.: "My brother, I congratulate you on having presented work which has passed inspection and entitles you to wages. But before you can be taught how to receive them, you must return to the room whence you came, and be duly prepared, and received in due and ancient form."

Junior Deacon conducts the candidate to the preparation room.

The Right Worshipful Master now usually orders a recess, which is done as follows:

R. W. M. (one rap): "The lodge will be at ease until the sound of the gavel in the East."

The Junior Warden places his little column upright on his pedestal, while the Senior Warden lays down his.

SECTION SECOND.

To prepare the candidate for the following ceremonies of the Mark Master's degree, the Junior Deacon proceeds as follows: . In the first section he has been deprived of all money, his coat off and shirt sleeves rolled up. In addition to that he is now deprived of his vest, necktie and collar, his right breast bared and a cable tow is put four times around his body; he also wears an apron as a Fellow Craft, being thus in a working costume. (See cut.)

Thus prepared, the Junior Deacon leads him to the door of the lodge, upon which he gives four knocks, by two's with a pause between each two. The four coils of the cable tow, and the four knocks allude to the fourth degree in Masonry, being that in which he is now initiated.

R. W. M. (one rap, seating the members, or, as it is said in the language of Masonry, "Calling on the lodge.")

S. D. (rising): "Right Worshipful Master, there is an alarm at the door of the preparation room."

R. W. M.: "Attend to that alarm."

The Senior Deacon steps to the west side of the altar, makes the due guard and sign, saluting the Right Worshipful Master, (See p. 13,) thence to the preparation room door, upon which he also gives four knocks by two's, opens the door and says:

S. D.: "Who comes here?"

J. D.: "Brother James Hunt, who has been regularly initiated, passed and raised to the sublime degree of a Master Mason, and now seeks further promotion in Masonry by being advanced to the honorary degree of a Mark Master."

(See Hand Book of Freemasonry for Dr. Hunt's previous degrees.)

S. D.: "Brother Hunt, is it of your own free will and accord that you make this request?"

Candidate: "It is."

S. D.: "Brother Junior Deacon, is he duly and truly prepared?"

J. D.: "He is."

S. D.: "Is he worthy and well qualified?"

J. D.: "He is."

S. D.: "Has he presented a satisfactory specimen of his work and made suitable proficiency in the preceeding degrees?"

J. D.: "He has."

S. D.: "By what further right or benefit does he expect to gain admission?"

J. D.: "By the benefit of the pass."

S. D.: "Has he the pass?"

J. D.: "He has it not, but I have it for him."

S. D.: "Advance and give it."

The Junior Deacon advances a few steps and whispers in the ear of the Senior Deacon the word, Joppa—the pass-word in this degree.

S. D.: "The pass is right. Let him wait until the Right Worshipful Master is informed of his request, and his answer returned."

The Senior Deacon closes the door, goes to the west side of the altar, and salutes the Right Worshipful Master with the due guard and sign, as in the accompanying figures.

Due-Guard, Mark Master.   Sign, Mark Master.

R. W. M.: "Brother Senior Deacon, what is the cause of the alarm?"

S. D.: "Right Worshipful Master, there is without Brother James Hunt, who has been regularly initiated, passed and raised to the sublime degree of a Master Mason, and now seeks further promotion in Masonry by being advanced to the honorary degree of Mark Master."

R. W. M.: "Is it of his own free will and accord that he makes this request?"

S. D.: "It is."

R. W. M.: "Is he duly and truly prepared,
worthy and well qualified?"

S. D.: "He is."

R. W. M.: "Has he presented a satisfactory
specimen of his work and made suitable proficiency
in the preceding degrees?"

S. D.: "He has."

R. W. M.: "By what further right or benefit
does he expect to gain admission?"

S. D.: "By the benefit of the pass."

R. W. M.: "Has he the pass?"

S. D.: "He has it not, but his conductor gave
it for him."

R. W. M.: "Give me the pass."

S. D. (with due guard and sign): "Joppa."

R. W. M.: "You will admit the candidate and
receive him in due form."

The Senior Deacon, saluting the Right Worshipful Master, as usual, returns to the door of the preparation room, which he opens without further
ceremony, and says:

S. D.: "It is the order of the Right Worshipful
Master that the candidate enter this lodge of Mark
Master Masons and be received in due form."

The Junior Deacon conducts the candidate into
the lodge and about three feet inside the door,
where he is halted by the Senior Deacon, who receives him in due form as follows:

S. D.: "Brother Hunt, I receive you into this lodge of Mark Masters upon the edge of the engraver's chisel against your naked right breast"—places the edge of the chisel against his breast—"under the pressure of the mallet"—strikes the head of the chisel lightly with the mallet—"which is to teach you that the moral precepts of this degree should make a deep and lasting impression upon your future life and conduct." (See figure.)

The Senior Deacon then, handing the chisel and mallet to the Junior Deacon, takes the candidate by the right arm and conducts him four times about the lodge room, walking rather slowly, and invariably.

following the course of the sun, and hence the circuits around the room always begin at the East. As they come in front of the stationed officers in their first circuit, the Junior Warden gives one rap with his gavel; the Senior Warden one, and the Right Worshipful Master one.

The second time around, the Junior Warden raps twice; the Senior Warden twice, and the Master twice.

The third time around the Junior Warden gives three raps—two rapidly then one (**　*) the Senior Warden the same, and the Right Worshipful Master the same.

During the fourth circuit the Junior Warden gives four raps—two then a pause, and then two more—the Senior Warden the same, (**　**) and the Master the same. (**　**)

As the Senior Deacon and candidate pass on around the lodge the Right Worshipful Master also reads from the Monitor as follows:

R. W. M. (1st circuit): "And I will cut wood out of Lebanon, as much as thou shalt need; and we will bring it to thee in floats, by sea to Joppa, and thou shalt carry it up to Jerusalem." 2 Chron. 2: 16. ("R. A. Standard," p. 18.)

R. W. M. (2nd circuit): "Then he brought me back the way of the gate of the outward sanctuary, which looketh toward the east, and it was shut. Then said the Lord unto me, This gate shall be shut, it shall not be opened, and no man shall enter in by it; because the Lord, the God of Israel, hath entered in by it; therefore it shall be shut."

R. W. M. (3rd circuit): "It is for the prince; the prince, he shall sit in it to eat bread before the Lord; he shall enter by the way of the porch of that gate, and shall go out by the way of the same."

R. W. M. (4th circuit): "And the Lord said unto me, Son of man, mark well, and behold with thine eyes and hear with thine ears, all that I say unto thee concerning all the ordinances of the house of the Lord, and all the laws thereof; and mark well the entering in of the house, with every going forth of the sanctuary." *(Ezk. 44: 1, 2, 3. 5, in Royal Arch Standard, p. 18.)

The Senior Deacon and candidate having thus made four complete circuits around the room, arrive in front of the Junior Warden's station, when the Senior Deacon gives four raps.

J. W. (rising): "Who comes here?"

S. D.: "Brother James Hunt, who has been regularly initiated, passed and raised to the sublime degree of a Master Mason, and now seeks further promotion in Masonry by being advanced to the honorary degree of a Mark Master."

J. W.: "Brother Hunt, is it of your own free will and accord that you make this request?"

Candidate: "It is."

J. W.: "Brother Senior Deacon, is the candidate duly and truly prepared, worthy and well qualified?"

S. D.: "He is."

---

* Mark well the blasphemy of using God's Word to build up a Godless, Christless, prayerless oath-bound system of Masonry as the Scriptures are used in these Chapter degrees.

J. W.: "Has he presented a satisfactory specimen of his work and made suitable proficiency in the preceding degrees?"

S. D. "He has."

J. W.: "By what further right or benefit does he expect to obtain this favor?"

S. D.: "By the benefit of the pass."

J. W.: "Has he the pass?"

S. D.: "He has it not, but I have it for him."

J. W.: "Advance and give me the pass."

The Senior Deacon whispers in his ear, Joppa, as at the door.

J. W.: "The pass is right. You will conduct him to the Senior Warden in the West for further examination."

Arriving at the Senior Warden's station, the Senior Deacon gives four knocks as before—two and two.

S. W. (rising): "Who comes here?"

S. D. "Brother James Hunt, who has been regularly initiated, passed and raised to the sublime degree of a Master Mason and now seeks further promotion in Masonry by being advanced to the honorary degree of a Mark Master," etc., the same dialogue as at the Junior Warden's station; after which the Senior Warden says:

S. W.: "The pass is right. You will conduct him to the Right Worshipful Master in the East for final examination."

Arriving in front of the Master's chair, the Senior Deacon knocks on the floor, two and two, as before.

R. W. M.: "Who comes here?"

S. D.: "Brother James Hunt, who has been regularly initiated, passed and raised to the sublime degree of a Master Mason, and now seeks further promotion in Masonry by being advanced to the honorary degree of a Mark Master."

R. W. M.: "Brother Hunt, is it of your own free will and accord you make this request?"

Candidate: "It is."

R. W. M.: "Brother Senior Deacon, is the candidate duly and truly prepared, worthy and well qualified?"

S. D.: "He is."

R. W. M.: "Has he presented a satisfactory specimen of his work and made suitable proficiency in the preceding degrees?"

S. D.: "He has."

R. W. M.: "By what further right or benefit does he expect to obtain this favor?"

S. D.: "By the benefit of the pass."

R. W. M.: "Has he the pass?"

S. D.: "He has it not but I have it for him."

R. W. M.: "Advance and give me the pass."

The pass, Joppa, is given in a whisper, as before.

R. W. M.: "The pass is right. You will re-conduct the brother to the West and place him in charge of the Senior Warden, who will teach him how to approach the East in a proper manner."

The Senior Deacon returns with the candidate to the west side of the altar, and about six feet from it, facing the Senior Warden's station, where he says:

S. D.: "Brother Senior Warden, it is the order of the Right Worshipful Master that you teach the candidate how to approach the East in a proper manner."

S. W.: "Brother Senior Deacon, face the candidate to the East." The Senior Deacon faces the candidate about and stands at his right.

S. W.: "Brother Hunt, you will now approach the East by four regular steps--first as an Entered Apprentice"—he takes the Entered Apprentice step; —"second, as a Fellow Craft"—he takes the Fellow Craft step;—"third, as a Master Mason;"—he takes that step also. "Now you will take one advance step with your right foot, bringing the heel of the left to the heel of the right, your feet forming a right angle."

The Senior Deacon sees to it that these orders are obeyed in a proper manner, by instructing the candidate whenever necessary.

S. W.: "Right Worshipful Master the candidate is before the altar."

R. W. M.: "My brother, for the fourth time you are before the altar of Freemasonry; but before you can proceed further, it is necessary that you take upon yourself a solemn obligation pertaining to this degree, which contains nothing that will conflict with the duty you owe to God, your country, your family, or yourself. With this assurance do you wish to proceed?"

Candidate: "I do."

R. W. M.: "Brother Senior Deacon, you will place the candidate at the altar in due form to be made a Mark Master."

The Senior Deacon leads the candidate to the altar and causes him to kneel on both knees facing the East, his body erect and both hands resting on the Holy Bible, square and compass, and reports:

S. D.: "Right Worshipful Master, the candidate is in due form." .

R W. M. Gives three raps, calling all the brethren to their feet. They move toward the altar and stand in two parallel lines, extending from the altar eastward, facing each other. The Senior Warden ought to stand at the west end of the north line, the Junior Warden at the west end of the south line, while the Senior Deacon stands immediately behind, and a little towards the right of the candidate. The Marshal stands at the east end of the south line, and the Junior Deacon at the east end of the north line—as in the accompanying diagram.*

The Right Worshipful Master steps to the East point of the altar, removes his hat, and addressing the candidate, says:

R. W. M.: "Say I, with your name in full, and repeat after me."

OBLIGATION

OF MARK MASTER MASON.

"I, James Hunt, of my own free will and accord, in the presence of Almighty God and this Right Worshipful Lodge of Mark Masters, erected to God

---

* In the annexed diagram E. represents East, or Master's station; W. is West, or Senior Warden's station; S. is South, or Junior Warden's station; R. W. M. is Right Worshipful Master; S. W. is Senior Warden; J. W., Junior Warden; S. D., Senior Deacon; J. D., Junior Deacon; M., Marshal; C., Candidate, and the black dots ● ● represent members. P. is preparation room; A. is ante-room.

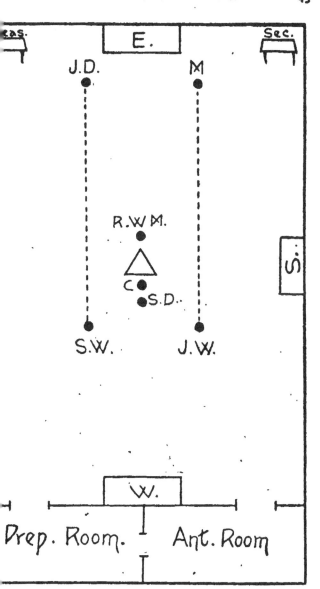

and dedicated to the memory of Hiram Abiff, do hereby and hereon sincerely promise and solemnly swear, that I will not reveal the secrets of this degree to a brother of a preceding degree, nor to any person in the world, except it be within a legally constituted and duly opened Mark Master's lodge, or to a brother of this degree whom I shall have found to be such by due trial, strict examination or legal information.

"I do furthermore promise and swear that I will answer and obey all due signs and summons sent to and received by me from a Mark Master's lodge, or given me by a brother of this degree, if within the length of my cable tow.

"I do furthermore promise and swear that I will receive the Mark of a brother Mark Master Mason when offered to me as a pledge, and will grant his request if in my power; if not, I will return his Mark with the price thereof, which is a Jewish half shekel of silver, equal in value to the fourth part of a dollar.

"I do furthermore promise and swear that I will not alter or change the Mark which I shall hereafter select after the same has been recorded in the lodge book of Marks.

"I do furthermore promise and swear that I will not loan or sell my Mark, nor pledge it a second time until it shall have been redeemed from its former pledge.

"I do furthermore promise and swear that I will not cheat, wrong or defraud a Mark Master's lodge, or a brother of this degree, out of the value of a

ay's wages, which is one penny, nor permit the
ame to be done by another, if in my power to
revent it.

"All this I sincerely promise and swear without
he least hesitation, equivocation or mental reserva-
ion, binding myself under no less a penalty than
hat of having my right ear smote off and my right
and struck off, should I violate this my solemn
bligation as a Mark Master Mason. So help me
od, and keep me steadfast."

R. W. M. (continuing): "In token of your sin-
erity you will detach your hands and kiss the holy
Bible now open before you."

R. W. M.: "The brother being now bound to
s by a tie that can never be broken, Brother Senior
Deacon, you will release him from the cable tow."

The Right Worshipful Master resumes his hat
nd steps back to the parallel lines of brethren,
acing the candidate, of course.

R. W. M. (addressing candidate): "Brother
Hunt, you now behold me approaching you from
he East with the step"—makes the step of a Mark
Master—"under the due-guard"—makes that sign as
n p. 36—"and penal sign of a Mark Master Mason"
—makes the penal sign, see p. 36—"and extending to
ou my right hand in token of continued friendship,
rotherly love and confidence."

The Master takes the candidate by the right
and, as in ordinary hand-shaking, holds it for an
nstant, and then releasing it says:

"But, my brother, before you arise from this
acred altar I desire to call your attention to one of

the ties of your obligation, wherein you have just solemnly sworn that you will receive the Mark of a brother Mark Master Mason when offered you as a pledge, and will grant his request if in your power; and if not, that you will return his Mark with the price thereof, which is a Jewish half-shekel of silver, equal in value to the fourth part of a dollar. I now request you to loan me — dollars (two, three, or four, as he pleases), for which I offer you my Mark as a pledge."

The Right Worshipful Master offers his Mark—same as the little keystone charm worn on the watch-

chain—to the candidate, who, of course cannot comply with the request, all his money having been removed in the preparation room. The candidate hesitates to take the Mark, in which case the Right Worshipful Master says:

R. W. M.: "You have sworn to receive a brother's Mark. You cannot refuse to take my Mark."

Candidate takes the Mark and the Master continues:

R. W. M: "Will you grant my request?"

Being unable to do so, his money having been taken from him in the preparation room, the candidate feels somewhat embarrassed, especially if he has never read an exposition of the chapter, such as Duncan's or Avery Allyn's.

R. W. M.: "Will you return my Mark so that I may apply to some other brother?"

The candidate probably offers to return the Mark, but the Right Worshipful Master, with an emphatic gesture, refuses to accept it, and says:

R. W. M: "You have sworn that you would return a brother's Mark with the price thereof, which is a Jewish half shekel of silver, equal in value to the fourth part of a dollar; you must return my Mark with its price."

If all this twaddle is new to the candidate, of course his embarrassment is greatly increased, but if he has posted himself beforehand, he simply pretends to be confused, and smiles inwardly at the credulity and silliness of those around him.

S. D.: "Right Worshipful Master, the brother cannot comply with that part of his obligation which binds him to return a brother's Mark with the price thereof; he is entirely destitute."

R. W. M: "Is it possible my brother, that you, after having assumed such obligations, and even before leaving the altar, are so unfortunate as to be compelled to violate them; examine yourself carefully and see if you cannot find about your person somewhere concealed so small a sum as the price of my Mark."

Candidate feels his clothes, searching here and and there, but of course finds nothing.

R. W. M.: "Do you find yourself entirely destitute?"

Candidate: "I do; yes, sir."

R. W. M.: "Brethren, you see before you a brother Mark Master Mason who is so entirely destitute that he is unable to comply with that part of his obligation which requires him to return a brother's Mark with the price. Who will assist him?"

All the brethren respond, "I will," and rush toward the candidate with money. He accepts a quarter or fifty-cent piece from one.

R. W M.: "This enables you, my brother, to return my Mark with its price." Takes the Mark and money from candidate. "This demand is made on you at this time, when on your bended knees at the altar, to impress upon your mind in the most solemn manner that you should never hastily reject the application of a worthy brother, especially when accompanied by so sacred a pledge as his Mark, but grant his request if in your power; if not, return him his Mark with the price thereof, which will enable him to procure the common necessities of life."

R. W. M. (taking candidate by the right hand): "I now raise you from a square to a perpendicular, and with the assistance of the Senior Deacon will instruct you in the grips and words of a Mark Master. Take me as I take you."

Pass Grip of a Mark Master.

He takes the candidate by the right hand, clasping his fingers with those of the candidate, as one would naturally do when assisting another up a steep place as in the annexed cut.

R. W. M.: "What is that?"

S. D. (replying): "The pass-grip of a Mark Master."

R. W. M.: "Has it a name?"

S. D.: "It has."

R. W. M.: "Will you give it to me?"

S. D.: "I did not so receive it, and cannot so impart it."

R. W. M.: "How will you dispose of it?"

S. D.: "I will syllable it with you."

R. W. M.: "Syllable it, and begin."

S. D.: "Begin you."

R. W. M.: "Nay, you must begin."

S. D.: "Jop."

R. W. M.: "pa."

S. D.: "Joppa."

R. W. M. (to candidate): "This, my brother, is the pass-grip of a Mark Master, the name of which is Joppa. It alludes to the ancient port of Joppa, where much of the material for the building of King Solomon's Temple was landed after being brought from Mount Lebanon by sea on floats. Masonic tradition informs us that the coast at that point was so steep that it was difficult for the workmen to ascend without assistance, which was afforded them by means of this strong grip given by others stationed there for the purpose; since which time this word and grip have been adopted as the pass to be given to gain admission into all well regulated Mark Masters' lodges."

R. W. M.: "Will you be off or from?"

S. D.: "From."

R. W. M.: "From what, and to what?"

S. D.: "From the pass-grip to the true grip of a Mark Master."

R. W. M.: "Pass." The Right Worshipful Master locks his little finger with that of the candidate, the backs of their hands being together. Each one now closes his fingers, the tops of their thumbs touching, as in the previous sign. The Senior Deacon always assists the candidate to place his fingers right. See figure below.

Real Grip of a Mark Master.

R. W. M.: "What is that?"

S. D.: "The true grip of a Mark Master."

R. W. M.: "Has it a name?"

S. D.: "It has."

R. W. M.: "Will you give it to me?"

S. D.: "I did not so receive it, and cannot so impart it."

R. W. M.: "How will you dispose of it?"

S. D.: "I will syllable it with you."

R. W. M.: "Syllable it, and begin."

S. D.: "Begin you."

R. W. M.: "Nay, you must begin."

S. D.: "Mark."

R. W. M.: "Well." Addressing candidate. "This is the true grip of a Mark Master, the name of which is Mark Well. It alludes to a certain text of Scripture: 'Then He brought me back the way of the gate of the outward sanctuary, which looketh toward the east, and it was shut. And the Lord said unto me, Son of man, Mark well, and behold with thine eyes, and hear with thine ears, all that I say unto thee concerning all the ordinances of the house of the Lord and all the laws thereof; and Mark well the entering in of the house with every going forth of the sanctuary.'" Ezekiel 44: 1-5. "R. A. Standard."

R. W. M (continuing): "Will you be off or from?"

S. D : "Off."

Master drops candidate's hand, returns to his station in the East, and seats the brethren with one rap of the gavel. The Senior Deacon and candidate remain standing at the west side of the altar.

R. W. M.: "Brother Senior Deacon, conduct the candidate to the East."

The candidate is conducted to the East, passing to the north of the altar, and is placed standing in front of the Right Worshipful Master's chair, who explains to him the Working Tools as follows:

R. W. M.: "My brother, I will now present you with the Working Tools of a Mark Master Mason, and will teach you their use."

"THE WORKING TOOLS OF A MARK MASTER MASON

are the Chisel and Mallet"—holds them in his hands. "The Chisel"—presents it to the candidate–"is an instrument made use of by operative masons to cut, carve, mark and indent their work. It morally demonstrates the advantages of discipline and education. The mind like the rough ashlar when taken from the quarry is rude and unpolished; but as the effect of the chisel in the hands of the skillful workman soon outlines and perfects the carved capital, the stately shaft, and the beautiful statue, so education discovers the latent virtues of the mind, and draws them forth to range the large field of matter and space, to display the summit of human knowledge, our duty to God and to man.

"THE MALLET

is used by operative masons to knock off excrescences and to smooth surfaces. It morally teaches to correct irregularities, and reduce man to a proper level; so that by quiet deportment he may, in the school of discipline, learn to be content. What the Mallet is to the workman, enlightened reason is to the passions; it curbs ambition, it depresses envy, it moderates anger, and it encourages good dispositions; whence arises among good Masons that comely order,

> "Which nothing earthly gives or can destroy,
> The soul's calm sunshine and the heartfelt joy."

R. W. M. (to candidate): "You will now be re-conducted to the place from whence you came, there be re-invested of what you were divested, and await my further will and pleasure."

R. W. M. (continuing): "The lodge will be at ease until the sound of the gavel in the East"—that is, a recess is ordered.

The Senior Deacon conducts the candidate to the west side of the altar, where he is taken in charge by the Junior Deacon and both retire to the preparation room, where he resumes his clothing and returns to the lodge. After awhile the Right Worshipful Master gives one rap, and the brethren are again seated, or, in the language of Masonry—"the lodge is called on."

The lodge is no sooner called to order than all stand up again, talk to one another in small groups, lounge about and act the part of idle men, remarking to one another that they don't know what to do next.

R. W. M. (gives one rap and inquires): "B,other Senior Warden, why this idleness of the Craftsmen; why are they not pursuing their labors?"

. S. W.: "Right Worshipful Master, the Temple is nearly completed, but the Craft are at a stand for the want of a Keystone belonging to one of the principal arches, which no one has received orders to make."

R W. M.: "That piece of work was assigned to our Grand Master, Hiram Abiff, and from his well-known punctuality I feel content that it was completed according to its original design prior to his death. Brother Overseers, you will repair to the East."

The three Overseers approach the R. W. M., who hands them a design of. the Keystone with its Mark, (circles and letters,) and asks:

R. W. M.: "Has a piece of work bearing that Mark been presented to you for inspection?"

The Overseers pretend to examine the design, and to consult together about it, when, after a minute or two the

M. O. replies: "Right Worshipful Master, on reflection we find there has, but it being neither oblong nor square, square work only being such as we had orders to receive, and not having the Mark of any of the Craft upon it, and we, not knowing the Mark which was upon it, were unanimous in concluding it unfit for use and heaved it over among the rubbish."

R. W. M.: "That is truly unfortunate, for no less depends on that stone than the completion of the Temple. You will therefore make strict search throughout the apartments of the Temple and among the rubbish, to see if it can be found."

The three Overseers pretend to search, looking here and there for a couple of minutes, and the Keystone is found near the Treasurer's desk, where the Junior Overseer placed it in a former part of the work. (See p. 25.) They then report.

M. O.: "Right Worshipful Master, strict search has been made and the stone is found."

R. W. M.: "Present it."

They present the stone to the Right Worshipful Master, who places it upon his pedestal, with the.

circles and letters in full view of the brethren. He
then gives one knock—the Overseers are seated.

R. W. M.: "Brother Senior Deacon, you will
conduct our newly admitted brother to the East."

The Senior Deacon conducts the candidate to a
seat facing the Right Worshipful Master in the
East and retires to his place.

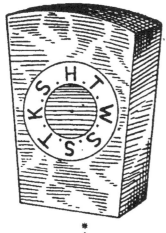

R. W. M., giving one knock on the stone with
his gavel, says: "This is the stone which was set at
naught of you builders, which is become the head of
the corner." Acts iv: 11.

★★

Gives two quick knocks, and reads: "Did you
never read in the Scriptures, The stone which the
builders rejected, the same is become the head of
the corner?" Matt. xxi: 42.

★★ ★

Gives three knocks—two, then one—and reads:
"And have you not read this Scripture: The stone

which the builders rejected is become the head of the corner?" Mark xii: 10.

** **

Gives four knocks—two and two—and reads: "What is this then that is written: The stone which the builders rejected, the same is become the head of the corner?" Luke xx: 17.

The R. W. M. then rehearses to the candidate "The Legend of the Keystone," explains the signs, etc., as follows:

R. W. M.: "My brother, as a preparatory circumstance attending your advancement to this degree, you were caused to represent one of the Craftsmen employed at the building of King Solomon's Temple, whose custom it was on the evening of the sixth day of every week to carry up their work for inspection, and, if approved, to receive wages.

"The work was inspected by three Overseers appointed by King Solomon for that purpose, and stationed at the south, west and east gates of the Temple. There were employed at the building of the Temple eighty thousand Craftsmen, and one would naturally suppose that in so great a number our Grand Masters were liable to be imposed upon, by unskillful Craftsmen presenting work unfit for use, but they were not; for King Solomon took the precaution that each Craftsman presenting work should have his Mark placed thereon, so that it might be easily known and distinguished when promiscuously brought up for inspection. * A brother's

---

* Allowing three minutes only for the inspection of each man's piece of work, and ten hours for a working day, it would take $\frac{80,000 \times 3}{60} = 4,000$ hours, and $\frac{4,000}{10} = 400$ working days, or a year and thirty-five days for the examination. How the labor of over a year could be crowded into a part of every Friday must be one of the Masonic secrets—Humbug.

Mark, therefore, became synonymous with his name, and hence he had no more right to alter or change his Mark than he had his name. The wages of a faithful Craftsman, when his work was approved, was one penny a day. In so great a number as 80,000 it would be natural to suppose that our Grand Masters were liable to be defrauded by Craftsmen demanding wages not their due; but they were not, for King Solomon took the further precaution that each Craftsman claiming wages should thrust his right hand through a lattice window into the apartments of the Senior Warden with a copy of his Mark in the palm thereof, at the same time giving this token."

The R. W. M. here stretches out his right hand towards the seated candidate with the two first fingers and the thumb extended and the two last fingers clinched. (See figure below.)

Receiving Wages.

"The amount of wages due that particular Mark was placed upon his fingers, with which he withdrew his hand, and so each brother passed on in succession until all were paid. * This token" (makes the sign again) "alludes to the way and manner in which each Craftsman was taught to receive his wages. Its use was to distinguish a true Craftsman from an imposter; and when an imposter was detected, the penalty was to have his right hand struck off. It was for the want of this token" (makes it again) "that you were detected and came near losing your right hand."

"I will now invest you with two additional signs to those given you at the altar. This, you will remember, is the due guard,"—makes that sign—"and this the sign of a Mark Master"—makes the sign.

Due Guard.   Sign.

* Allowing but one minute for each man, it would take 80,000 minutes, or 1,333 hours and 20 minutes, or 133 days, 3 hours and 20 minutes to pay the Craft; and 400 days to inspect the work, makes over 533 days. or over two and one-half years to be taken out of every week. Mystery! Superlative Mystery!!

"This is the grand hailing sign, or sign of distress, of a Mark Master Mason, and alludes to the way and manner in which the Craft were taught to carry their work for inspection." The R. W. M. makes the sign by extending the two first fingers and thumb of the right hand, the two last fingers clinched, the arm outstretched. It is also the sign of receiving wages. (See figure p. 59.)

"This is the principal sign, and alludes to the principal words, which are heave over"—makes the sign by the motion of throwing something over his left shoulder with both hands. (See figure.) "They

Heave Over.

allude to the rejection of the Keystone by the three Overseers, which circumstance happened as follows:

"After our Grand Master, Hiram Abiff, had finished that piece of work, but before he had given orders to have it carried up to the Temple, he was assassinated, as we have had an account in the preceding degree.

"It so happened on the sixth day of a certain week, when the Craft went to carry their work for inspection, that a young Fellow Craft, seeing that stone and supposing it designed for some part of the building, took it up with him. On presenting it to the Junior Overseer at the South gate he remarked that it was neither oblong nor square, neither had it the Mark of any of the Craft upon it, but, owing to its singular form and beauty, he was unwilling to reject it, and directed him to present it to the Senior Overseer at the West gate for further inspection. He, for the same reason, directed him to present it to the Master Overseer at the East gate for final inspection. On presenting it to the Master Overseer at the East gate, he called together his brother Overseers for consultation, remarking that it was neither oblong nor square, neither had it the Mark of any of the Craft, and they not knowing the Mark that was upon it, concluded it unfit for use and agreed to heave it over among the rubbish.

"Soon after, the Senior Grand Warden informed King Solomon that the Temple was nearly completed, but that the Craft were at a stand for the want of a keystone belonging to one of the principal arches, which no one had received orders to furnish. King Solomon, being well satisfied that our Grand Master, Hiram Abiff, had furnished that piece of work according to the original design prior to his assassination, made inquiry of the Overseers if a piece of work bearing a certain Mark had been presented to them for inspection?

"After consultation they replied there had, but it being neither oblong nor square, and not having the Mark of any of the Craft upon it, and they not knowing that which was upon it, concluded it unfit for use and agreed to heave it over among the rubbish. King Solomon then ordered a strict search to be made through the apartments of the Temple and among the rubbish to see if it could be found; strict search was accordingly made, the stone found, and afterwards applied to its intended use.

"Its color was white, and to it * alludes a certain text of Scripture, which says: 'To him that overcometh will I give to eat of the hidden manna, and I will give him a white stone, and in the stone a new name written, which no man knoweth saving he that receiveth it.' (Rev. 2: 17.) The New Name is composed of the words of which the letters on the stone are the initials, namely: Hiram, The Widow's Son, Sent To King Solomon. In undergoing an examination you will alternate the words in this manner." The words are given alternately by the Senior Deacon and R. W. M., the former beginning with the first word, as follows:

S. D.: "Hiram."

R. W. M.: "The."

S. D.: "Widow's."

R. W. M.: "Son,"

S. D.: "Sent."

* This is simply brutal blasphemy, and how any minister of the Gospel of God, or even a professing Christian, can support such iniquity can be explained only on the ground of 2 Thes. 2: 11, 12.

R. W. M.: "To."

S. D.: "King."

R. W. M.: "Solomon."

Continuing the Right Worshipful Master says:

"This was, originally the Mark of our Grand Master Hiram Abiff"—pointing to the circle of letters—"now the general Mark of this degree; in the center of which each brother places his own particular Mark to which the obligation alludes. Any design you may choose to select placed within this circle of letters constitutes your particular Mark, which you have sworn you will not alter or change after you have once chosen it, and it has been recorded as such in the lodge book of Marks. And I would impress upon you, my brother, that it is your duty to promptly select your Mark and have it duly recorded in the book of Marks.

"This degree was founded by our three ancient Grand Masters, Solomon, King of Israel, Hiram, King of Tyre, and Hiram Abiff, to be conferred upon all those who were found worthy and well qualified; not so much as an honorary reward for their zeal, fidelity and attachment to Masonry as to render it impossible for a brother who was found worthy to be advanced to this degree to fall into such extreme destitution as to suffer for the common necessaries of life when the price of his Mark would procure the same.

"A brother presenting his Mark and claiming assistance represents our Grand Master Hiram Abiff, who was a poor man, but from his upright and regular conduct, and his great skill in architecture and the sciences, became greatly distinguished among the Craft.

"A brother receiving a Mark, and granting assistance, represents our Grand Master King Solomon, who was a rich man, and from his wealth and great liberality became exceedingly distinguished."

R. W. M.: "Brother Hunt, you will now arise, and I will deliver you the charge pertaining to this degree."

ADDRESS TO THE CANDIDATE. *

R. W. M.: "My brother, I congratulate you on having been thought worthy of being advanced to this honorary degree of Masonry. Permit me to impress it on your mind, that your assiduity should ever be commensurate with your duties, which become more and more extensive as you advance in Masonry. In the honorable character of Mark Master Mason it is more particularly your duty to let your conduct in the world, as well as in the lodge, and among your brethren be such as may stand the test of the Great Overseer's square, that you may not, like the unfinished and imperfect work of the negligent and unfaithful of former times, be rejected and thrown aside, as unfit for that spiritual building—that house not made with hands—eternal in the heavens.

"While such is your conduct, should misfortunes assail you, should friends forsake you, should envy traduce your good name, and malice persecute you, yet may you have confidence that among Mark Mas-

* Royal Arch Standard, 1897. p. 23.

ter Masons you will find friends who will administer relief to your distresses and comfort your afflictions, ever bearing in mind, as a consolation under all the frowns of fortune, and as an encouragement to hope for better prospects, that the stone which the builders rejected, (possessing merits to them unknown) became the chief stone of the corner."

R. W. M.: "Brother Senior Deacon, you will now conduct our newly admitted brother to a seat."

R. W. M.: "Brother Junior Warden, what is the hour?"

J. W. (rising): "Right Worshipful Master, it is the sixth hour of the sixth day of the week."

R. W. M.: "This is the day and the hour when the Craft should repair to the apartment of the Senior Warden to receive their wages. You will give your orders accordingly."

J. W. (gives three knocks): "Brother Master of Ceremonies, you will assemble the Craft for the purpose of repairing to the apartments of the Senior Warden to receive their wages."

Master of Ceremonies: "Brethren, you will form in procession on the north side of the lodge, single file, facing the East."

The procession is formed on the north side of the lodge, facing East, as directed by the Master of Ceremonies. The Overseers are in front, Master, Senior and Junior Overseers in their order; then the candidate, and then the brethren.

The procession moves around the lodge room twice, going by way of the East toward the West,

and as they march they sing the first four verses of
the Mark Masters' song, as follows:

Music—America.

"Mark Masters, all appear
Before the Chief O'erseer,
   In concert move.
Let him your work inspect,
For the Chief Architect;
If there be no defect,
   He will approve.

"You who have passed the square,
For your rewards prepare,
   Join heart and hand;
Each with his mark in view,
March with the just and true;
Wages to you are due,
   At your command.

"Hiram, the widow's son,
Sent unto Solomon
   Our great keystone;
On it appears the name
Which raises high the fame
Of all to whom the same
   Is truly known.

"Now to the westward move,
Where full of strength and love,
   Hiram doth stand;
But if imposters are
Mix'd with the worthy there,
Caution them to beware
   Of the right hand.

Passing around the lodge room the second time
each brother thrusts his right hand through the
Lattice Window at the West, giving the proper token

The march and song are so timed that the fourth verse is finished as the last man is paid his wages. See cut p. 59.

All the Craft having thus been paid, they "compare notes," and finding upon quiet inquiry that the candidate also has received a penny, there is much dissatisfaction and loud murmuring expressed as they stand in front of the Senior Warden's station.

R. W. M.: "Brother Senior Warden, what is the cause of this confusion?"

S. W.: "Right Worshipful Master, the Craft are not satisfied with their wages."

R. W. M.: "Have you not paid every man according to agreement?"

S. W.: "I have."

R. W. M.: "Then, brethren, why are you dissatisfied?"

Master of Ceremonies: "Right Worshipful Master, we who have borne the burden and heat of the day complain that those who came in at the eleventh hour have been made equal unto us."

R. W. M.: "Will you hear the law?"

M. of C.: "We will."

R. W. M.: "Not only hear it, but abide by it?"

M. of C.: "We will, but we know of no such law."

R. W. M.: "I will read you the law." Reads from the Manual the parable in Matt. xx: 1-16, the brethren standing grouped in front of the Senior Warden's station.

"It is like unto a man that is an householder, which went out early in the morning to hire laborers

into his vineyard. And when he had agreed with
the laborers for a penny a day, he sent them into his
vineyard. And he went out about the third hour,
and saw others standing idle in the market-place,
and said unto them, Go ye also into the vineyard,
and whatsoever is right I will give you. And they
went their way. Again he went out about the sixth
and ninth hour, and did likewise. And about the
eleventh hour, he went out and found others standing idle, and saith unto them, Why stand ye here all
the day idle? They say unto him, Because no man
hath hired us. He saith unto them, Go ye also into
the vineyard, and whatsoever is right that shall ye
receive. So when even was come, the lord of the
vineyard saith unto his steward, Call the laborers,
and give them their hire, beginning from the last
unto the first. And when they came that were hired
about the eleventh hour, they received every man a
penny. But when the first came, they supposed that
they should have received more; and they likewise
received every man a penny. And when they had
received it, they murmured against the good man of
the house, saying, These last have wrought but one
hour, and thou hast made them equal unto us, which
have borne the burden and heat of the day. But he
answered one of them and said, Friend, I do thee no
wrong; didst not thou agree with me for a penny?
Take that thine is, and go thy way; I will give unto
this last, even as unto thee. Is it not lawful for me
to do what I will with mine own? Is thine eye evil
because I am good? So the last shall be first, and
the first last." (R. A. Standard.)

Having finished reading, the R. W. M. says:

R. W. M.: "What say you now, are you satisfied?"

M. of C.: "We are."

The brethren, still standing, sing the last verse of the song, and the R. W. M. returns to the East.

"Now to the praise of those
Who triumphed o'er the foes
    Of Mason's art;
To the praiseworthy three,
Who founded this degree;
May all their virtues be
    Deep in our hearts."

R. W. M. gives one rap and the brethren are seated.

CLOSING CEREMONIES.

R. W. M.: "Brother Junior Deacon, what is the last as well as the first great care of Masons when in lodge assembled?"

J. D.: "To see that the lodge is duly tyled, Right Worshipful."

R. W. M.: "Perform that duty; and inform the Tyler that I am about to close this lodge of Mark Master Masons, and direct him to tyle accordingly."

J. D. gives four knocks by two's, as is always done in this degree.

Tyler knocks in the same manner.

J. D. opens the door, informs the Tyler, closes the door again and says: "That duty is performed, and the lodge is duly tyled, Right Worshipful."

R. W. M.: "How tyled, brother Junior Deacon?"

J. D.: "By, a worthy brother without, armed with the proper implement of his office."

R. W. M.: "What are his duties there?"

J. D.: "To keep off cowans and eavesdroppers, and admit none but such as are duly qualified and have permission from the Right Worshipful Master."

R. W. M. (one rap, and Junior Deacon is seated): "Brother Senior Warden, are you a Mark Master Mason?"

S. W.: "I am; try me."

R. W. M.: "How will you be tried?"

S. W.: "By the chisel and mallet."

R. W. M.: "Why by the chisel and mallet?"

S. W.. "Because they are the working tools of a Mark Master."

R. W. M.: "Where were you made a Mark Master?"

S. W.: "In a legally constituted and duly opened lodge of Mark Masters."

R. W. M.: "How many compose such a lodge?"

S. W.: "Eight or more."

R. W. M.: "When composed of eight who are they?"

S. W.: "The Right Worshipful Master, Senior and Junior Wardens, Senior and Junior Deacons and the Master, Senior and Junior Overseers."

Here follow the same questions and answers precisely as in opening the lodge down to the words "give them proper instruction for their labor." After which:

R. W. M. (gives three raps and all the brethren stand): "Brother Senior Warden it is my order that this lodge of Mark Masters be now closed; this order you will communicate to the Junior Warden in the South and he to the brethren present for their government."

S. W.: "Brother Junior Warden, it is the order of the Right Worshipful Master that this Mark Masters' lodge be now closed. This order you will communicate to the brethren present for their government.".

J. W.: "Brethren, it is the order of the Right Worshipful Master that this Mark Masters' lodge be now closed; take due notice thereof and govern yourselves accordingly. Look to the East."

Following the lead of the Right Worshipful Master they now make all the signs from the Mark to that of the Entered Apprentice degree inclusive as represented on p. 12, 13; * after which:

R. W. M. gives two raps, Senior Warden two raps, Junior Warden two.

R. W. M. (again) two raps, S. W. two raps, J. W. two.

The Right Worshipful Master or Chaplain then reads or mumbles the following alleged prayer, and if desirable, music may be introduced either before or after.

"Supreme Grand Architect of the Universe, who sitteth on the throne of mercy, deign to view our labors in the cause of virtue and humanity with the eye of compassion; purify our hearts, and cause us to know and serve thee aright. Guide us in the

---

* N. B. In opening the lodge the signs were made from the Entered Apprentice up to Mark Master inclusive. In closing, the signs are made from Mark Master down to Entered Apprentice, and so in all the degrees

paths of rectitude and honor; correct our errors by the unerring square of thy wisdom, and enable us to practice the precepts of Masonry, that all our actions may be acceptable in thy sight. Amen."

Brethren: "So mote it be."

R. W. M.: "Accordingly I declare this lodge of Mark Masters closed. Brother Junior Deacon, so inform the Tyler. Brother Senior Deacon, take charge of the three great lights."

All remove their aprons and jewels of office, the Tyler opens the ante-room door and they usually repair to some convenient sample room, generally by invitation of the candidate, thus going from labor to refreshment.

CHAPTER III.

A Mark Master Mason, desiring to receive the next degree, must commit the following lecture carefully to memory, as he cannot be advanced, or at least ought not to be, unless he has "made suitable proficiency in the preceding degree." He is examined, therefore, in an open Mark Masters' lodge, the examination being conducted by the Right Worshipful Master, Senior Deacon, or any other well-posted brother.

If a Chapter Mason—a Mark Master for instance —wishes to visit a strange lodge of Mark Masters, unless he can be vouched for, he must in like manner pass a thorough examination in the following lecture, and hence it is of the utmost importance not only that this lecture be absolutely correct, but that it be thoroughly memorized.

Many Chapter Masons, however, possibly the majority of them, are densely ignorant as to the degrees through which they have passed, and take but little interest in lodge meetings beyond the Royal Arch degree. They attend the meetings of the Chapter on the occasions of a banquet, or when some prominent man—a popular preacher, or a leading politician—is the candidate, and some who are ambitious to hold office in the various lodges will also be prompt in their attendance, but the rank and file of the members, once they have obtained the privilege of dangling a little keystone from their watch chains, simply content themselves with being known as Chapter Masons, and when an ordinary lodge meeting takes place, are conspicuous by their absence. I know several Chapter Masons here in Chicago who received two, and some of them even three degrees in one night, and who, as a result, have but very little knowledge of Chapter Masonry.

I also know quite a number, and have heard of several others, who, by a careful study of the "Hand Book of Freemasonry," have visited the Blue Lodge as often as they pleased. though never initiated, passed, or raised in a regular lodge, and I want to emphasize the assertion, that any intelligent young man, who carefully studies out the degrees as given in this book, and who thoroughly memorizes the lectures as herein set forth can readily visit any of the lodges if he wishes to do so.

THE LECTURE.

MARK MASTERS' DEGREE-- FIRST SECTION.

Examiner: "Are you a Mark Master?"

Candidate or Visitor: "I am; try me."

Ex.: "How will you be tried?"

Can.: "By the chisel and mallet."

Ex.: "Why by the chisel and mallet?"

Can.: "Because they are the working tools of a Mark Master."

Ex.: "What were some of the preliminary steps attending your advancement to the degree of Mark Master?"

Can.: "I was caused to represent one of the Craft at the building of King Solomon's Temple, whose custom it was on the sixth day of each week, to carry up their work for inspection."

Ex.: "Who inspected their work?"

Can.: "Three Overseers, appointed by King Solomon, and stationed at the South, West and East gates."

Ex.: "How many Fellow Crafts were engaged at the building of King Solomon's Temple?"

Can.: "Eighty thousand."

Ex.: "Among so large a number were not our Grand Masters liable to be imposed upon by unworthy Craftsmen presenting work unfit for use?"

Can.: "They were not, because King Solomon took the precaution to cause each Craftsman to have a Mark which he was ordered to place upon his work, that it might be distinguished when brought up to the Temple."

Ex. "What were the wages of a Fellow Craft?"

Can.: "A penny a day."

Ex. "Among so large a number were not our Grand Masters liable to be imposed upon by unworthy Craftsmen demanding wages when none were their due."

Can.: "They were not, because King Solomon took the further precaution to cause each Craftsman applying for wages to thrust his right hand through a Lattice Window into the apartments of the Senior Warden with a copy of his Mark in the palm thereof, at the same time giving this token"—holding out his right hand, palm upward, the two first fingers and thumb extended, the other two fingers clinched as in the figure below.

Ex.: "To what does that token allude?"

Can.: "To the manner in which a Fellow Craft received his wages."

SECOND SECTION.

Ex.: "Where were you made a Mark Master?"

Can.: "In a legally constituted and duly opened Mark Masters' lodge."

Ex.: "Where were you prepared?"

Can.: "In a room adjoining the same."

Ex.: "How were you prepared?"

Can.: "By being deprived of all money, divested of my outward apparel, in a working costume, with a cable tow four times about my body, in which condition I was conducted to the door of the lodge and caused a regular alarm to be made by four distinct knocks."

Ex.: "To what do those four knocks allude?"

Can.: "To the fourth degree of Masonry, it being that upon which I was about to enter."

Ex.: "What was said to you from within?"

Can.: "Who comes here?"

Ex.: "Your answer?"

Can.: "A worthy brother who has been regularly initiated, passed and raised to the sublime degree of a Master Mason, and now seeks further promotion in Masonry by being advanced to the honorary degree of a Mark Master."

Ex.: "What was then asked you?"

Can.: "If it was of my own free will and accord I made the request; if I was duly and truly prepared, worthy and well qualified; if I had exhibited a satisfactory specimen of my skill, and made suitable proficiency in the preceding degrees to entitle me to this. All of which being answered in the affirmative, I was then asked by what particular right or benefit I expected to gain admission."

Ex.: "Your answer?"

Can.: "By the benefit of a pass-word."

Ex.: "Had you the pass?"

Can.: "I had it not; my conductor gave it for me."

Ex.: "What was it?"

Can.: "Joppa."

Ex.: "To what does it allude?"

Can.: "To the ancient port of Joppa, where much of the material for the building of King Solomon's Temple was landed after being brought from Mount Lebanon by sea on floats."

Ex.: "What was then said to you?"

Can.: "I was told to wait until the Right Worshipful Master could be informed of my request and his answer returned."

Ex.: "What was his answer?"

Can.: "Let the candidate enter and be received in due form."

Ex.: "How were you received?"

Can.: "Upon the edge of the engraver's chisel, under the pressure of the mallet; which was to teach me that the moral precepts of this degree should make a deep and lasting impression upon my future life and conduct." (See figure p. 38.)

Ex.: "How were you then disposed of?"

Can.: "I was conducted four times regularly about the lodge to the Junior Warden in the South, the Senior Warden in the West, and the Right Worshipful Master in the East, where the same questions were asked and answers were returned as at the door."

Ex.: "How did the Right Worshipful Master dispose of you?"

Can.: "He ordered me to be re-conducted to the West and placed in charge of the Senior Warden, who would teach me how to approach the East in a proper manner."

Ex.: "What was that proper manner?"

Can.: "Advancing by four upright steps; first as an Entered Apprentice, second as a Fellow Craft, third as a Master Mason; fourth, advancing one step with my right foot, bringing the heel of my left to the heel of my right, my feet forming a right angle, my body erect, facing the East."

Ex.: "What was then done with you?"

Can.: "I was made a Mark Master Mason."

Ex.: "How?"

Can.: "In due form."

Ex.: "What was that due form?"

Can.: "Kneeling at the altar on both knees, both hands resting on the Holy Bible, Square and Compass; in which position I took upon myself the solemn oath, or obligation of a Mark Master."

Ex.: "Have you that oath, or obligation?"

Can.: "I have." (See p. 44.)

Ex.: "Repeat it." (It is generally repeated.)

Ex.: "After your obligation, what followed?"

Can.: "Being bound by a tie that can never be broken, I was ordered released from the cable tow."

Ex.: "What did you next behold?"

Can. "The Right Worshipful Master approaching me from the East with the step, due guard and penal sign of a Mark Master Mason, who extended to me his right hand in token of continued brotherly love and confidence, raised me from a square to a perpendicular, and presented me with the pass, token of the pass, grip and word of a Mark Master."

Ex.: "How were you then disposed of?"

Can.: "I was ordered to be conducted to the East, where the Right Worshipful Master presented me with the working tools of a Mark Master, and taught me their uses."

Ex.: "What are the Working Tools of a Mark Master?"

Can.: "The Chisel and Mallet."

Ex.: "What is the use of the Chisel?"

Can.: "The Chisel is an instrument used by operative masons to cut, carve, mark and indent their work."

Ex.: "What does it morally teach?"

Can.: "The Chisel morally demonstrates the advantages of discipline and education. The mind, like the rough ashler, when taken from the quarry, is rude and unpolished; but as the effect of the chisel in the hands of the skillful workman soon outlines and perfects the carved capital, the stately shaft, and the beautiful statue, so education discovers the latent virtues of the mind, and draws them forth to range the large field of matter and space, to display the summit of human knowledge, our duty to God and man."

Ex.: "What is the use of the Mallet?"

Can.: "The Mallet is an instrument used by operative masons to knock off excrescences and to smooth surfaces."

Ex.: "What does it morally teach?"

Can.: "It morally teaches to correct irregularities, and reduce man to a proper level; so that, by quiet deportment, he may, in the school of discipline, learn to be content. What the mallet is to the workman, enlightened reason is to the passions; it curbs ambition, it depresses envy, it moderates anger, and it encourages good dispositions; whence arises among Masons that comely order,

"Which nothing earthly gives, or can destroy,
  The soul's calm sunshine, and the heart-felt joy."

Ex.: "What followed?"

Can.: "I was ordered to be re-conducted to the place from whence I came and be there invested of what I had been divested and await the further will and pleasure of the Right Worshipful Master."

Ex.: "Have you any signs belonging to this degree?"

Can.: "I have."

Ex.: "Show me a sign."

Candidate makes the Grand Hailing Sign of a Mark Master.

Ex.: "What is that?"

Can.: "The grand hailing sign, or sign of distress, of a Mark Master Mason." (See figure p. 77)

Ex.: "To what does it allude?"

Can.: "To the manner in which the Craft were taught to carry their work."

Ex.: "Show me another sign."

Candidate makes the "heave over" sign as below.

Ex.: "To what does that allude?"

Can.: "To the rejection of the keystone by the Overseers."

Ex.: "How happened that circumstance?"

Can.: "After our Grand Master Hiram Abiff had finished that piece of work, but before he had given orders to have it carried up to the Temple, he was assassinated, as we have an account in a preceding degree. It so happened on the sixth day of a certain week when the Craft went to carry their work for inspection, that a young Fellow Craft seeing that stone and supposing it designed for some part of the building took it up with him. On presenting it to the Junior Overseer at the South gate he remarked that it was neither oblong nor square, neither had it the mark of any of the Craft upon it, but owing to its singular form and beauty he was unwilling to reject it, and directed him to present it to the Senior Overseer at the West gate for further inspection; he for the same reason directed him to present it to the Master Overseer at the East gate for final inspection. On presenting it to the Master Overseer at the East gate, he called together his brother Overseers for consultation, remarking that it was neither oblong nor square, neither had it the mark of any of the Craft upon it, and they not knowing the mark that was upon it, concluded it unfit for use and agreed to heave it over among the rubbish.'

Ex.: "What followed?"

Can.: "Soon after King Solomon was informed that the Craft were at a stand for want of a keystone belonging to one of the principal arches which no one had received orders to furnish '

Ex.: "What followed?"

Can.: "King Solomon well satisfied that our Grand Master, Hiram Abiff, had furnished that piece of work according to the original design prior to his assassination, made inquiry of the Overseers if a piece of work bearing a certain mark had been presented to them for inspection? After consultation they replied that there had, but it being neither oblong nor square, and not having the mark of any of the Craft upon it, and they not knowing that which was upon it, concluded it unfit for use and heaved it over among the rubbish."

Ex.: "What followed?" ·

Can.: "King Solomon ordered strict search to be made among the rubbish of the Temple to see if it could not be found; search was made, the stone found and devoted to its intended use."

Ex.: "Give me another sign."

Candidate makes the "due guard." (See figure below.)

Due-Guard Mark Master.

Ex.: "What is that?"

Can.: "The due guard of a Mark Master."

Ex.: "To what does it allude?"

Can.: "To the penalty of my obligation."

Ex.: "Show me another sign."

Candidate gives the "penal sign" as in annexed figure:

Sign, Mark Master

Ex.: "To what does it allude?"

Can : "To the additional penalty of my obligation."

Ex.: "Give me the 'pass grip' of this degree."

Candidate gives the pass grip. (See figure.)

Pass Grip of a Mark Master.

Ex.: "Has it a name?"

Can.: "It has."

Ex. "Will you give it to me?"

Can.: "I did not so receive it, and cannot so impart it."

Ex.: "How will you dispose of it?"

Can.: "I will syllable it with you."

Ex.: "Syllable it, and begin."

Can.: "Jop."

Ex.: "Pa."

Can.: "Joppa."

Ex.: "To what does it allude?"

Can.: "To the ancient port of Joppa, where much of the material for the building of King Solomon's Temple was landed, after being brought from Mount Lebanon by sea on floats."

Ex.: "Will you be off or from?"

Can.: "From."

Ex.: "From what and to what?"

Can.: "From the pass-grip of a Mark Master to the true grip of the same."

Ex.: "Pass."

Candidate links the little finger of his right hand with that of his Examiner, both hands being back to back; hands are then closed, the points of both thumbs touching. (See figure.)

Real Grip of a Mark Master.

Ex.: "What is that?"

Can. "The true grip of a Mark Master"

Ex.: "Has it a name?"

Can.: "It has."

Ex.: "Will you give it to me?"

Can.: "I did not so receive it, and cannot so impart it."

Ex.: "How will you dispose of it?"

Can.: "I will syllable it with you."

Ex.: "Syllable it, and begin."

Can.: "Mark."

Ex.: "Well."

Can.: "Mark well."

It often happens that in alternating this password, the one examined gives "Well," and the Examiner "Mark," the one examined then repeating "Mark Well." Joppa is often alternated in the same manner.

Ex.: "On what was this degree founded?"

Can.: "On the keystone of a certain arch in King Solomon's Temple."

Ex.: "By whom was it wrought?"

Can.: "By our Grand Master, Hiram Abiff."

Ex.: "What was its color?"

Can.: "White, and to it alludes a certain text of Scripture, which says: 'To him that overcometh will I give to eat of the hidden manna, and I will give him a white stone, and in the stone a new name written, which no man knoweth saving he that receiveth it.'"

Ex.: "What was that new name?"

Can.: "The words of which the letters on that stone are the initials."

Ex.: "What are those words?"

Can.: "Hiram."

Ex.: "The."

Can.: Widow's.

Ex.: "Son."

Can.: "Sent."

Ex.: "To."

Can.: "King."

Ex.: "Solomon."

Ex.: "By whom was this degree founded?"

Can.: "By our three ancient Grand Masters, Solomon, King of Israel, Hiram, King of Tyre, and Hiram Abiff."

Ex.: "For what purpose?"

Can.: "To be conferred upon all those who were found worthy and well qualified; not so much as an honorary reward for their zeal, fidelity and attachment to Masonry, as to render it impossible for a brother, who was found worthy to be advanced to this degree, to fall into such extreme destitution as to suffer for the common necessities of life, when the price of his Mark would procure the same."

Ex.: "Were you required at any time during your advancement to this degree, to comply with the tie of your obligation?"

Can.: "I was."

Ex.: "At what time?"

Can.: "When at the altar on my bended knees."

Ex.: "Why at that time?"

Can.: "To impress upon my mind in the most solemn manner that I should never hastily reject the application of a worthy brother for assistance. especially when accompanied by so sacred a pledge as his Mark, but grant his request if in my power; if not, to return his Mark with the price thereof, which would enable him to procure the common necessaries of life."

Ex.: "Whom does a brother represent presenting his Mark and claiming assistance?"

Can.: "Our Grand Master, Hiram Abiff, who was a poor man, but from his upright and regular conduct, and his great skill in architecture and the sciences, became greatly distinguished among the Craft."

Ex.: "Whom does a brother represent receiving a Mark and granting assistance?"
Can.: "Our Grand Master, King Solomon, who was a rich man, and from his wealth and great liberality became exceedingly distinguished."

Ex.: "What followed?"

Can.: "The Right Worshipful Master delivered me the charge of a Mark Master." (See R. A. Standard. p 23.)

This ends the Lecture of the Mark Master's degree and in concluding this third chapter I desire to call the reader's attention to a few very important points in connection with the initiatory ceremonies.

First: In opening the lodge a portion of Scripture is read from which the name of Jesus Christ is knowingly, and deliberately, expunged; but to make this still clearer, I shall here place the Scripture as quoted by Masonry, and the Scripture as contained in God's Word in juxtaposition.

MASONRY.	GOD'S WORD.
Charge to be read at opening the lodge.	Wherefore laying aside all malice, and all guile, and hypocrisies, and envies and all evil speakings,
"Wherefore, brethren, lay aside all malice, and guile, and hypocrisies, and envies, and all evil speakings.	2. As new-born babes, desire the sincere milk of the word, that ye may grow thereby:
"If so be, ye have tasted that the Lord is gracious, to whom, coming as unto a living stone, disallowed, indeed, of men, but chosen of God, and precious; ye, also, as living stones be ye built up a spiritual house, an holy priesthood, to offer up sacrifices acceptable to God."	3. If so be ye have tasted that the Lord is gracious.
	4. To whom coming, as unto a living stone, disallowed indeed of men, but chosen of God. and precious.
	5. Ye also, as living stones, are built up a spiritual house, an holy priesthood, to offer up spiritual sacrifices, acceptable to God by Jesus Christ.—1 Pet ii: 1-5.

This willful mutilation of God's Word and the absolute rejection of the Lord Jesus Christ is an unvarying principle in Freemasonry, as all intelligent Masons know, and every candidate, minister or layman, Mohammedan or Methodist, is solemnly sworn under a death penalty to "ever maintain and support" this God-dishonoring principle. Thus rejecting Jesus Christ Masonry rejects the true God—Father, Son and Holy Spirit—and hence it is a damning delusion for Masons to make their boast of "trust in God," while they knowingly repudiate and dishonor God, and do not believe a single statement He has made. Another question: How can Freemasons expect to dwell with Jesus Christ in heaven, when they so impiously reject and black ball Him from their lodges on earth?

He has positively declared, "He that denieth me before men (as in the Masonic lodge) shall be denied before the angels of God." (Luke xii: 9.) Masons do you believe this? Masonic preachers do *you* believe it?

Second: The assertion made in the preceding Lecture with such solemn assurance, that the glorious promise of the Lord Jesus to the Overcomer contained in Rev. ii: 17 refers directly to the Keystone of the Mark Master's degree, and that the "new name" which Christ will confer upon His faithful people, has reference to the name of Hiram Abiff—"Hiram The Widow's Son Sent To King Solomon." (see p. 87) is a blasphemy more glaring and brutal than any expression that has ever fallen from the lips of an Ingersoll or a Tom Paine. This every professing Christian, and above and beyond all, every preacher ought to know, and yet these miserable men for a few paltry loaves and fishes will "solemnly promise and swear" to "ever maintain and support" this horrible mockery of truth.

All Masonic teachers are unanimous in declaring that "Search of truth is the great object of all Masonic teaching," and that "the keystone of a Mark Master is therefore the symbol of a fraternal covenant among those who are engaged in the common search after Divine Truth." ("Masonic Ritualist;" Mackey, p. 288.) What a burlesque upon intelligence and moral honesty is this! Pretending to be in search of Divine Truth, and at the same time willfully rejecting and even hating Him who is "the Way, the Truth and the Life." (John xiv: 6) and

without whom none can come to the Father. Surely
the words of the apostle may be fittingly applied to
the Masonic fraternity in general, and especially to
Masonic preachers: "The god of this world hath
blinded the minds of them which believe not, lest
the light of the Gospel of the glory of Christ who is
the image of God should dawn upon them." (2 Cor.
iv: 4, R. V.)

CHAPTER IV.

PAST MASTER'S DEGREE.

The term Past Master is applied in Freemasonry to a Master Mason who has presided for one or more terms as Worshipful Master of a Masonic lodge. It is an official title, carrying with it even the privilege of a voice in the councils of the Grand Lodge, and renders one eligible to any position in the gift of the Craft, even that of Grand Master.

In Chapter Masonry there is a degree called the "Past Master's degree," but in that case, the term "Past Master" has an entirely different meaning; it confers no Masonic dignity, and is simply nothing more than a qualification for receiving the Royal Arch degree. The term of office in a Masonic lodge is one year, in all other alleged secret societies it is only six months. The Master Mason who has presided over his lodge for one or more terms is called an actual Past Master, while a Mason who has merely received the Past Master's degree in the Chapter is termed a Virtual Past Master. In a regularly constituted lodge, no one is elected to the office of Master, or installed as such, unless he has shown superior skill and intelligence, both as to Masonic law and usage and the working of the degrees, and upon him is conferred the degree of Past Master, or, as it is called, "The secrets of the chair" at his installation; but although a man may have received the degree of Past Master in the Chapter, yet he cannot take any part in conferring "the secrets of the chair" upon the Master-elect, and neither can an "Actual" Past Master be allowed to sit in a Past Master's lodge unless he has actually taken that degree in the Chapter.

The following is what "The Royal Arch Standard," the latest authorized Masonic Monitor, has to say concerning this Past Master's degree that I am now about to illustrate:

"This degree is more closely connected with Symbolic than with Chapter Masonry. Those who receive the degree in the Chapter are termed 'virtual' Past Masters, in contradistinction to those who have been elected and installed in a regularly constituted symbolic lodge, who are called 'actual' Past Masters, the former having no rights or privileges as such out of the Chapter.

"The officers of a Past Master's lodge are the same as those of a lodge of Master Masons; the officers in the Chapter ranking as follows, namely: The High Priest as Master; the King as Senior Warden; the Scribe as Junior Warden; the Treasurer and Secretary occupy corresponding stations; the Captain of the Host as Marshal; the Principal Sojourner as Senior Deacon; the Royal Arch Captain as Junior Deacon, and the Tyler at his proper station. "The jewel of a Past Master is a pair of compasses extended to sixty degrees, the points resting on the segment of a circle. Between the extended legs of the compasses is a flaming sun." (See Hand Book of Freemasonry, frontispiece.)

From the preceding pages it will be seen that the candidate receiving the Mark Master's degree is said to be "advanced," while in this he is said to have "Presided as Master in the chair" or "Regularly passed the chair;" and in conferring the degree the aprons and jewels of the Chapter are used as in that of Mark Master.

OPENING CEREMONIES.

A lodge of Past Masters is convened simply for the purpose of conferring the degree, and like that of Mark Master is entirely under the control of the Chapter, where the Candidate is balloted for, and where he is examined in the lecture. The first thing done at every lodge meeting is for the brethren to put on their aprons, and that is sometimes attended to as much as fifteen or twenty minutes before the Master's gavel sounds in the East.

᛫ The usual hour of opening the lodge having arrived, and the brethren and officers being clothed in their aprons and jewels of office, the aprons worn in the Royal Arch degree—the Right Worshipful Master gives one rap with his gavel, and all are seated. The lodge is now to all intents and purposes a lodge of Master Masons, and the first order the Right Worshipful gives is:

R. W. M.: "Brother Junior Deacon, you will see that the Tyler is at his post, and close the door."

J. D. Looking into the anteroom, and closing the door, he replies without saluting, "The Tyler is at his post, Right Worshipful."

R. W. M. (one rap): "Brother Senior Warden, are you satisfied that all present are Past Master Masons?"

Casting his eyes around the lodge room and noticing that no strangers are present, he at once replies:

S. W.: "All present are Past Master Masons. Right Worshipful."

But should he see any stranger in the room he answers:

S. W.: "I will ascertain by the proper officers and report." (Continuing): "Brothers Senior and Junior Deacons approach the West and give me the pass of a Past Master."

The two Deacons meet at the west side of the altar and walk briskly towards the Senior Warden, into whose ear they whisper the word, Giblim.

S. W.: "You will now proceed upon my right and left, collect the pass from the brethren and convey it to the West."

The Deacons, as in the preceding degree, pass along in front of the brethren, each brother rising and whispering the word—Giblim—in the Deacon's ear; and having thus collected the pass from all the brethren, except the Master and Junior Warden, they return to the West and report:

S D.: "All are Past Masters in the North, brother Senior Warden."

J. D.: "All are Past Masters in the South."

S. W.: "All present are Past Masters, Right Worshipful."

No one is allowed to remain, not even an actual Past Master, though in possession of the password,

Giblim, unless he has received this degree in the Chapter.

R. W. M.: "Brother Junior Deacon, what is the first great care of assembled Masons?"

J. D.: "To see the lodge duly tyled."

R. W. M.: "Perform that duty, and inform the Tyler that we are about to open a lodge of Past Masters, and direct him to tyle accordingly."

The Junior Deacon opens the door, informs the Tyler as the Master had directed, then closes it and gives five raps—two, and three—which are answered in the same manner by the Tyler outside; then facing the East and making the due guard of a Past Master as in the accompanying figure, he reports:

J. D.: "That duty is performed, Right Worshipful, and the lodge is duly tyled."

R. W. M.: "How is it tyled?"

J. D.: "By a worthy brother without, armed with the proper implement of his office." .

R. W. M.: "What are his duties there?"

J. D.: "To keep off all cowans and eavesdroppers, and admit none but such as are duly qualified and have permission from the Right Worshipful Master."

R. W. M. Gives one rap and J. D. is seated.

R. W. M.: "Brother Senior Warden, are you a Past Master?"

S. W. (rising): "I have the honor so to be."

R. W. M.: "How gained you that important distinction?"

S. W.: "By having been duly elected to preside over a legally constituted lodge of Free and Accepted Masons."

R. W. M.: "How shall I know you to be a Past Master?"

S. W.: "From a grip to a span, from a span to a grip."

R. W. M.: "Where were you made a Past Master?"

S. W.: "In a legally constituted and duly opened lodge of Past Masters."

R. W. M.: "How many compose such a lodge?"

S. W.: "Three, or more."

R. W. M.: "When composed of only three, who are they?"

S. W.: "The Right Worshipful Master, Senior and Junior Wardens."   .

R. W. M.: "The Junior Warden's station."

S. W.: . "In the South."

R. W. M. Gives two raps; the Junior Warden stands: "Why in the South, and your duties there, brother Junior Warden?"

J. W.: "As the sun in the South at meridian height is the glory and beauty of the day, so is the Junior Warden in the South, the better to observe the time, to call the Craft from labor to refreshment, superintend them during the hours thereof, see that none convert the purposes of refreshment into intemperance or excess, and call them on at the will and pleasure of the Right Worshipful Master."

R. W. M.: "The Senior Warden's station?"

J. W.: "In the West, Right Worshipful."

R. W. M.: ";Why in the West, and your duties there, brother Senior Warden?"

S. W.: "As the sun is in the West, at the close of the day, so is the Senior Warden in the West, to assist the Right Worshipful Master in opening and closing the lodge, to pay the Craft their wages, if any be due, and see that none go away dissatisfied; harmony being the strength and support of all well-governed institutions."   :

R. W. M: "The Right Worshipful Master's station?"

S. W.: "In the East."

R. W. M.: "Why in the East, and his duties there?"

S. W.: "As the sun rises in the East to open and govern the day, so is the Right Worshipful Master in the East to open and govern the lodge, set the Craft to work, and give them proper instruction for their labor."

R. W. M. (rising) Gives three raps; all the brethren stand. "Brother Senior Warden, it is my order that a lodge of Past Masters be now opened for the dispatch of business under the usual Masonic restrictions; this order you will communicate to the Junior Warden in the South, and he to the brethren for their government."

S. W.: "Brother Junior Warden, it is the order of the Right Worshipful Master that a lodge of Past Masters be now opened for the dispatch of business under the usual Masonic restrictions; this order you will communicate to the brethren for their government."

J. W.: "Brethren, it is the order of the Right Worshipful Master that a lodge of Past Masters be now opened for the dispatch of business under the usual Masonic restrictions; of this you will take due notice and govern yourselves accordingly. Look to the East."

The Right Worshipful Master, Wardens and brethren, make all the signs in concert, from the Entered Apprentice degree to that of Past Master inclusive. (For the signs of the preceding degrees see pages 12 and 13.)

R. W. M. Gives two raps—the S. W. two—the J. W. two.

R. W. M. Again two raps—the S. W. two—the J. W. two.

R. W. M. Then gives one rap—the S. W. one—the J. W. one.

After which the Right Worshipful Master, or a preacher if present, repeats or reads from the Monitor what is ignorantly called a prayer, addressed to the G. A. O. T. U—"The Great Architect of the Universe"—the old pagan title of Baal or the sun-god, and from which of course the name of Jesus Christ must be rigorously excluded.

OPENING PRAYER.

"Most holy and glorious Lord God, the Great Architect of the Universe, the Giver of all good gifts and graces: Thou hast promised, that where two or three are gathered together in thy name, thou wilt be in their midst, and bless them. In thy name we assemble, most humbly beseeching thee to bless us in all our undertakings, that we may know and serve thee aright, and that all our actions may tend to thy glory, and to our advancement in knowledge and virtue. And we beseech thee, O Lord God, to bless our present assembling, and to illuminate our minds, that we may walk in the light of thy countenance; and, when the trials of our probationary state are over, be admitted into THE TEMPLE 'not made with hands, eternal in the heavens'. Amen."

R. W. M.: "I now declare this lodge of Past Masters, erected to God and dedicated to the memory of the Holy SS. John, opened in due form. Brother Junior Deacon, inform the Tyler. Brother Senior Deacon, display the great light."

The Junior Deacon turns to the ante-room door and gives five raps—two, two, one (** ** *), which are answered by the Tyler in the same manner The J. D. then opens the door and informs the Tyler that the lodge is open; closes the door, and again gives the five raps as before, which are again answered by the Tyler. He now faces the East, makes the due guard, and reports: "The Tyler is informed, Right Worshipful."

In the meantime the Senior Deacon opens the Bible on the altar and places upon it the square and compass with the points extended and both above the square.

R. W. M.: Gives one rap and the lodge is seated, as in the diagram on page 104.

The reader is referred to pages 12 and 13 for the signs from the Entered Apprentice degree to that of Mark Master inclusive; the due guard and sign of a Past Master are made as follows:

For the due guard, close the fingers of the right hand, the thumb being extended upward. Place the top of the thumb thus extended against the center ot the lips closed, and draw it down and back under the lower jaw, as if motioning to "split the tongue from tip to root," alluding to the penalty of the Mark Master degree. (See annexed figure.)

For the penal sign draw the open palm of the right hand edgewise across the body from the left side of the neck diagonally to the right hip—the thumb and first finger being towards the body—alluding to the infliction of the penalties of the pre-

ceding degrees—"throat cut," "left breast torn open,"
and "body severed in twain."   (See annexed figure.)

Due Guard.   Sign.

"The General Grand Chapter of the United
States, in 1856, adopted a resolution recommending
the Chapters under its jurisdiction to abridge the
ceremonies now conferred in the Past Master's degree within the narrowest constitutional limits, only
retaining the inducting of the candidate into the
Oriental Chair, and communicating the means of
recognition."—R. A. Standard, p. 30.

Notwithstanding this recommendation, the sub-
ordinate Chapters throughout the country still con-

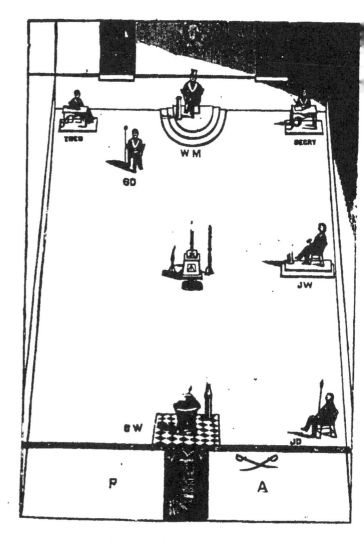

PLAN OF LODGE ROOM.

Degrees of Entered Apprentice, Fellow Craft and Master Mason.

tinued to indulge in almost unlimited buffoonery in conferring this degree, and it is only very recently that the ritual has been revised by the General Grand Chapter, so that now the initiatory ceremonies are generally confined within proper limits of propriety and decency.

I received this degree of Past Master twice, first in the Chapter, on which occasion all kinds of stupid horse-play were indulged in at my expense, and secondly in the Preparation room of Keystone Lodge No. 639, Chicago, in being installed as Worshipful Master of the lodge, and at which time, Joseph H. Dixon, the retiring Master of Keystone, Jno. O'Neil, Past Master of Blair lodge 393, Frank Holcomb, Past Master of Garden City 141, and the Masters and Past Masters of several other of the Chicago lodges took part in the simple ceremonies then performed. Following is the correct secret work as authorized by the General Grand Royal Chapter of the United States.

SECRET WORK.

R. W. M.: "Brethren, this lodge of Past Masters has been called and opened for the purpose of inducting into the Oriental Chair Brother James Hunt, who has been duly elected"—he was elected in the Chapter—"to that distinguished honor. If there are no objections, we will now proceed to the ceremonies of installation." He pauses for a second or two, and no objection being made, continues:

R. W. M.: "Brother Senior Deacon, you will present the Candidate."

There is no preparation whatever in this degree —no hoodwink, no cable tow, but the Candidate is simply clothed as a Master Mason—that is, he wears a Master Mason's apron.

The Senior Deacon, without knocking, leads the Candidate into the lodge and conducts him at once to the west side of the altar, where he presents him to the Right Worshipful Master as follows:

S. D.: "Right Worshipful Master, I have the pleasure of presenting to you Brother James Hunt, who has taken all the degrees of Freemasonry, from Entered Apprentice to Mark Master, inclusive, and now prays to be inducted into the Oriental Chair of King Solomon, having been duly elected to that distinguished honor."

R. W. M.: "My brother, permit me to congratulate you upon being elected to this honorable position; but before you can be installed it is necessary that you take a solemn obligation, pertaining to this degree, which contains nothing that will conflict with the duty you owe to God, your country, your family, or yourself. With this assurance, do you wish to proceed?"

Candidate: "I do."

R. W. M. Gives three raps, and all the brethren stand: "You will kneel on both knees at the altar, place both hands upon the Holy Bible, Square and Compass, pronounce your name in full, and repeat after me."

The Right Worshipful Master stands on the East of the altar, the Senior Deacon immediately behind and towards the right hand of the Candidate.

The Master removes his hat and administers the following

OBLIGATION,

which, as in all the degrees, the Candidate repeats, clause by clause, after the Master.

R. W. M.: "I, James Hunt, of my own free will and accord, in the presence of Almighty God and this worshipful lodge of Past Masters, erected to God and dedicated to the memory of the Holy Saints John, do hereby and hereon sincerely promise and solemnly swear, that I will not reveal the secrets of this degree to a brother of a preceding degree, nor to any person in the world except it be within a legally constituted and duly opened Past Masters' lodge, or to a brother of this degree whom I shall have found to be such by due trial, strict examination, or legal information.

"I do further promise and swear that I will not open or close this or any other lodge of Free and Accepted Masons over which I may be called to preside without giving a lecture, or a part of a lecture, or causing the same to be done.

"I do further promise and swear that I will not rule this or any other lodge of Free and Accepted Masons over which I may be called to preside in an arbitrary or unconstitutional manner, but will govern the same according to the ancient customs, usages and landmarks of the Craft.

"All this I sincerely promise and solemnly swear without the least hesitation, equivocation, or mental reservation, binding myself under no less a penalty than that of having my tongue split from tip to root, together with the infliction of the penalties of the preceding degrees, should I violate this my solemn obligation as a Past Master. So help me God, and keep me steadfast."

R. W. M.: "In token of your sincerity, you will kiss the Holy Bible." *

The Master resumes his hat and stepping back about four or five feet continues:

R. W. M.: "You now behold me approaching you from the East on the step, under the due guard and sign of a Past Master Mason."

Master takes the step and makes these signs as follows:

R. W. M.: "A Past Master steps off one step with the left foot, bringing the heel of the right to the toe of the left, his feet forming the right angle of a perfect square. The due guard is given thus: Clinch the fingers of the right hand, place the point of the thumb, extended vertically, at the center of the lips, then draw it downward and backward to the throat. It alludes to the penalty of your obligation, wherein you have sworn that you would suffer your tongue to be split from tip to root should you violate any portion of your Past Master's obligation. The sign of this degree is given by drawing the right hand edgewise—palm extended, and thumb downward—from the left side of the neck, diagonally across the body to the left side, thus covering all

* If Masonry is a good institution, if it is that "which is most needed in this age," if it is doing the work of the church, and a hand-maid of Christianity, then why impose such terrible oaths, and bind the Candidate under such horrible death penalties to keep it a profound secret?

the former penalties, in allusion to the remaining portion of the penalty of your obligation. On entering or retiring from a lodge of Past Masters, you will step to the altar and salute the Right Worshipful Master with this due guard and sign. In token of the continuance of my brotherly love and confidence, I now extend to you my right hand"—taking the Candidate by the grip of a Master Mason (figure I.); "arise."

R. W. M. (holding Candidate's hand): "Will be off or from?"

S. D. (for Candidate): "From."

fig. 1

R. W. M.: "From what and to what?"

S. D.: "From the grip of a Master Mason **to** the grip of a Past Master."

R. W. M.: "Pass."

S. D.: "From a grip to a span." The Master slips his right hand up and takes hold of the Candidate's right forearm above the wrist; the Candidate, guided by the Senior Deacon, takes hold of the Master's arm in the same manner. (See figure 2.)

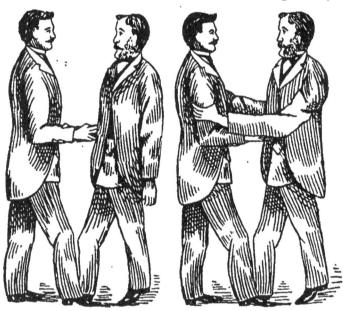

Fig. 2.   Fig. 3.

S. D.: "From a span to a grip." The Master now with his left hand takes the Candidate by the right arm, about midway between the elbow and shoulder, the Candidate, instructed by the Senior

Deacon, catches the Master's right arm with his left hand in the same way. (See figure 3.)

S. D.: "A two-fold cord is strong but a three-fold cord is not easily broken."·

The Master holding Candidate's fore arm above the wrist forms one cord, (as in figure 2,) grasping his arm between the elbow and shoulder forms Masonically "a two-fold cord," and the Candidate grasping the Master's arm above the elbow forms the Masonic "three-fold cord," (as in figure 3.)

During the examination of a Candidate, or of a brother wishing to visit a strange lodge of Past Masters, the grip is given as follows, and is introduced here so that the reader may have a clear understanding of how to give it.

The Right Worshipful Master, or Examiner, holding the Candidate by the Master Mason's grip, says:

R. W. M.: "Will you be off or from?"

Can.: "From."

R. W. M.: "From what, and to what?"

Can.: "From the grip of a Master Mason to the grip of a Past Master."

R. W. M.: "Pass."·

Can.: "From a grip to a span"—taking Master's fore arm.

R. W. M.: "From a span to a grip"—taking Candidate above the elbow.

Can.: "A two-fold cord is strong."

R. W. M.: "A three-fold cord is not easily broken.".

R. W. M. (at the altar): "What is that?"

S. D (for Candidate): "The grip of a Past Master."

R. W. M.: "Has it a name?"

S. D.: "It has."

R. W. M.: "Will you give it to me?"

S. D.: "I did not so receive it, and cannot so impart it."

R. W. M.: "How will you dispose of it?"

S. D.: "I will syllable it with you."

R. W. M.: "Syllable, and begin."

S. D.: "Gib."

R. W. M.: "Lim."

S. D.: "Giblim."

R. W. M.: "This, my brother, is the grip of a Past Master, and its name is Giblim, signifying Stone-squarer."

R. W. M.: "Will you be off or from?"

S. D.: 'Off."

The Right Worshipful Master drops the Candidate's hand, retires to his station in the East, and seats the lodge by one rap of his gavel. Senior Deacon and Candidate remain standing near the altar.

R. W. M.: "Brother Senior Deacon, you will conduct the Candidate to the East."

He is conducted in front of the Master's chair.

R. W. M.: "My brother, I now invest you with the jewel and badge of your office" (a square attached to a blue ribbon, is suspended from his neck) "and with pleasure conduct you to the Oriental Chair."

The Candidate is conducted to the chair, before which he stands, ready to be seated.

R. W. M.: "Masonic tradition informs us that when King Solomon became infirm he was assisted in taking his seat and rising therefrom in this manner by two Giblimites who attended him for that purpose."

The Master takes the Candidate by the Past Master's grip, as already explained, the Senior Deacon taking him by the left arm in the same manner.

R. W. M.: "From a grip to a span—from a span to a grip—a two-fold cord is strong—a three-fold cord is not easily broken."

The Candidate is seated in the Master's chair, as if he were old and infirm, and again assisted to rise in the same manner, and seated a second time.

R. W. M. (continuing): "As King Solomon wore a crown as an emblem of royal dignity, so as a mark of distinction, and agreeably to an ancient custom, you as Master are to be covered when presiding."

The Candidate places his hat on his head.

R. W. M.: "The gavel is an emblem of power and of your authority." He is handed the gavel. "One knock with the gavel calls the lodge to order, and should always be promptly obeyed when given. Calling an officer's name and giving one knock, causes him, only, to rise, thus: 'Brother Junior Deacon' (one knock); the J. D. alone rises. When up, one knock seats him. (The J D. sits.) Two knocks call all the officers to their feet," (Master gives two knocks and all the officers rise;) "one knock seats them." (Gives one knock and they are seated again.) "Three knocks cause all the officers and brethren to rise, and when standing one knock seats them." These knocks are given and promptly obeyed.

"The duties of the chair are many and various, and to the Master are confided, for his guidance and direction, all the implements of Masonry and the various furniture of the lodge, especially that great light in Masonry which will guide you to all truth, direct your path to the Temple of happiness, and point out to you the whole duty of man.

"My brother, you have been inducted into the Oriental Chair in due and ancient form: you have been intrusted with the emblem of authority, and instructed in some of the essential duties of the office. Long experience, and a careful study of the laws, customs and usages of Freemasonry, a patient and cheerful regard for the rights of every brother, and a firm and alert watchfulness to preserve the dignity and decorum of the brotherhood over whom you preside, are necessary to enable you to properly discharge the duties of your high office. At this time no particular test of your proficiency will be required; you will therefore now surrender your jewel and gavel, and take your place in front of the East."

The Candidate hands the gavel and the square to the Right Worshipful Master, lays his hat to one side, and stands in front of the Master's pedestal, where the Right Worshipful Master delivers the

ADDRESS TO THE CANDIDATE.

R. W. M.: "My brother: The Past Master's degree, unlike all the other degrees of Freemasonry, sheds no light upon itself. It was formerly conferred only on Masters of lodges, to instruct them in the duties they owed the lodges over which they were called to preside, and likewise the duties of the brethren to.the Chair; but we, as Royal Arch Masons, confer this degree, not only as a preliminary step. but also for the more important purpose of guarding us against a breach of our Masonic obligations.

"The conferring, at this time, of a degree which has no historical connection with the other capitular degrees is an apparent anomaly, which, however, is indebted for its existence to the following circumstances:

"Originally when Royal Arch Masonry was under the government of Symbolic lodges, in which the Royal Arch degree was then always conferred, it was a regulation that no one could receive it unless he had previously presided as the Master of that or some other lodge, and this restriction was made because the Royal Arch was deemed too important a degree to be conferred only on Master Masons. But, as by confining the Royal Arch to those only who had been actually elected as the presiding officers of their lodges, the extension of the degree would have been materially circumscribed, and its usefulness greatly impaired, the Grand Master often granted upon due petition his dispensation to permit certain Master Masons (although not elected to preside over their lodges) 'to pass the Chair,' which was a technical term, intended to designate a brief ceremony, by which the candidate was invested with the mysteries of a Past Master, and like him entitled to advance in Masonry as far as the Royal Arch, or the perfection and consummation of the third degree.

"When, however, the control of the Royal Arch was taken from the Symbolic lodges and intrusted to a distinct organization—that, namely, of Chapters —the regulation continued to be observed, for it was doubtful to many if it could be legally abolished, and, as the law still requires that the august degree of Royal Arch shall be restricted to Past Masters, our candidates are made to pass the Chair simply as a preparation and qualification towards being invested with the solemn instructions of the Royal Arch.

"Your receiving this degree confers upon you no official rank outside of the Chapter. The honors and peculiar privileges belonging to the Chair of Symbolic lodges are confined exclusively to those who have been 'duly elected to preside over and govern' such lodges, and who have been called 'Actual Past Masters,' whereas, those who receive the degree in the Chapter are termed 'Virtual Past Masters,' for, although they are invested with the secrets of the degree, yet they are not entitled to the rights and prerogatives of 'Actual Past Masters.'

"With this brief explanation of the reason why this degree is now conferred upon you, and why you have been permitted to 'preside as Master in the Chair,' you will retire and suffer yourself to be prepared for those further and profounder researches into Masonry which can only be consummated in the Royal Arch degree.

CLOSING THE LODGE.

R. W. M.: "Brother Junior Deacon, what is the last, as well as the first, great care of assembled Masons?"

J. D.: "To see the lodge duly tyled."

R. W. M.: "Perform that duty, and inform the Tyler that I am about to close this lodge of Past Master Masons: direct him to take due notice and tyle accordingly."

J. D. Gives three knocks, and then two, on the ante-room door, which are answered by three and two knocks by the Tyler.

J. D. opens the door, informs the Tyler, closes it again, and knocks as before, and the Tyler replies by similar knocks.

J. D Faces the East, makes the due guard—p. 103.: "That duty is performed, Right Worshipful, and the lodge is duly tyled."

R. W. M.: "How tyled, brother Junior Deacon?"

J. D.: "By a worthy brother without, armed with the proper implement of his office."

R. W. M.: "His duty there?"

J. D.: "To keep off all cowans and eavesdroppers, and admit none but such. as are duly qualified and have permission."

R. W. M. (one rap seats the J. D.): "Brother Senior Warden, are you a Past Master?"

S. W.: "I have the honor so to be."

R. W. M.: "How gained you that important distinction?"

S. W.: "By having been duly elected to preside over a legally constituted lodge of Free and Accepted Masons."

R. W. M.: "How shall I know you to be a Past Master?"

S. W.: "From a grip to a span, from a span to a grip, a two-fold cord is strong, but a three-fold cord is not easily broken."

R. W. M.: "Where were you made a Past Master?"

S. W.: "In a legally constituted and duly opened lodge of Past Masters."

R. W. M.: "How many compose such a lodge?"

S. W.: "Three, or more."

R. W. M.: "When composed of only three, who are they?"

S. W.: "The Right Worshipful Master, Senior and Junior Wardens."

R. W. M.: "The Junior Warden's station?"

S. W.: "In the South, Right Worshipful."

R. W. M. (two raps): "Why in the South, brother Junior Warden, and your duty there?"

J. W.: "As the sun in the South at meridian height is the glory and beauty of the day, so is the Junior Warden in the South, the better to observe the time, to call the Craft from labor to refreshment, superintend them during the hours thereof, see that none convert the purposes of refreshment into intemperance or excess, and call them on at the will and pleasure of the Right Worshipful Master"

R. W. M.: "The Senior Warden's station?"

J. W.: "In the West."

R. W. M.: "Why in the West, and your duty there, brother Senior Warden?"

S. W.: "As the sun is in the West at the close of the day, so is the Senior Warden in the West to assist the Right Worshipful Master in opening and closing the lodge, to pay the Craft their wages, if aught be due, and see that none go away dissatisfied; harmony being the strength and support of all well-governed institutions."

R. W. M.: "The Right Worshipful Master's station?"

S. W.: "In the East.".

R. W. M.: "Why in the East, and his duty there?"

S. W.: "As the sun rises in the East to open and govern the day, so is the Right Worshipful Master in the East, to open and govern the lodge, to set the Craft to work and give them proper instruction for their labor."

R. W. M. (gives three raps—all the brethren stand): "Brother Senior Warden, it is my order that this lodge of Past Masters be now closed; this order you will communicate to the Junior Warden in the South and he to the brethren present for their government."

S. W.: "Brother Junior Warden, it is the order of the Right Worshipful Master that this lodge of Past Masters be now closed. This order you will communicate to the brethren for their government."

J. W : "Brethren, it is the order of the Right Worshipful Master that this lodge of Past Masters be now closed; take due notice thereof and govern yourselves accordingly."

R. W. M.: "Together, brethren." They all then, in concert with the R. W. M., make the signs, from the Past Masters, down to the Entered Apprentice degree, inclusive. (See pages 12, 13, 103.)

The R. W. M. gives two knocks—Senior Warden, two—Junior Warden, two.

R. W. M. gives two again—the S. W., two—and J. W., two.

R. W. M. gives one knock—S. W., one, and J. W., one.

Thus each of the three principal officers gives five raps corresponding to the number of the degree, and the same at opening the lodge.

R. W. M.: "Accordingly I declare this lodge of Past Masters closed in due form. Brother Junior Deacon, so inform the Tyler. Brother Senior Deacon, take charge of the Great Lights." (Gives one rap.)

The Junior Deacon opens the door, and the Senior Deacon stepping to the altar picks up the square and compass and closes the Bible. It is the usual custom to confer the degrees of Past Master and Most Excellent Master upon a candidate the same night, and hence preparations are now made for the latter degree, during which there is a recess and the Candidate is requested to await either down stairs or in some other room in the building.

CHAPTER V.

In this honorary degree of Past Master I have included the opening, initiatory and closing ceremonies in one chapter. I shall now proceed to give the Lecture of this degree in a separate division; and shall suppose that a brother is being examined in the ante-room in order to visit a lodge of Past Masters.

THE LECTURE OF PAST MASTER.

Examiner: "Are you a Past Master?"

Visitor: "I have the honor so to be."

Ex.: "How gained you that distinguished honor?"

Vis.: "By having been duly elected and installed to preside over a legally constituted lodge of Free and Accepted Masons."

Ex.: "How shall I know you to be a Past Master?"

Vis.: "From a grip to a span, from a span to a grip."

Ex.: "Where were you made a Past Master?"

Vis.: "In a legally constituted and duly opened lodge of Past Masters."

Ex.: "How were you admitted?"

Vis.: "I was clothed as a Master Mason, conducted into the lodge without ceremony by the Senior Deacon, who placed me at the altar and there presented me to the Right Worshipful Master in the East."

Ex.: "What did the Right Worshipful Master then do with you?"

Vis.: "He made me a Past Master in due form."

Ex.: "What is that due form?"

Vis.: "Kneeling on both my knees, my body erect, and with both hands resting on the Holy Bible. Square and Compass, in which due form I took the oath or obligation of a Past Master."

Ex.: "Have you that oath or obligation?"

Vis.: "I have."

Ex.: "Repeat it."

Vis.: "I, (full name), of my own free will and accord," etc. See p. 107.

Ex.: "After your obligation what did you behold?"

Vis.: "The Right Worshipful Master approaching me from the East with the step, due guard and sign of a Past Master, and who, in token of the continuance of his brotherly love and confidence, extended to me his right hand and bid me to arise."

Ex. "What followed?"

Vis.: "I was conducted to the East, where the Right Worshipful Master invested me with the jewel of a Past Master, being the badge of my office, and he then conducted me to the Oriental Chair."

Ex.: "What followed?"

Vis.: "The Right Worshipful Master took me by the right hand—from a grip to a span—from a span to a grip—a two-fold cord is strong, but a three-fold cord is not easily broken; and with the assist-

ance of the Senior Deacon, inducted me into the
Oriental Chair, according to ancient custom."

Ex.: "What followed?"

Vis.: "I was informed that according to Masonic tradition that was the manner in which our
ancient Grand Master, King Solomon, when he had

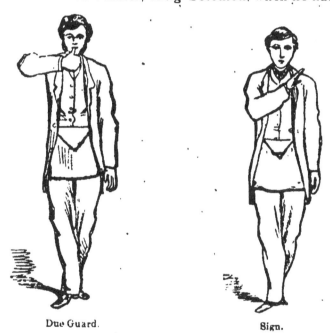

Due Guard.   Sign.

become old and infirm, was assisted in taking his
seat and rising therefrom by two Giblimites who attended him for that purpose "

Ex.: "What followed."

Vis.: "I was presented with the gavel, the emblem of power and of my authority, and instructed
in its use."

Ex.: "Have you any signs belonging to this degree?"

Vis.: "I have."

Ex.: "Give me a sign."

Visitor makes the due guard. (See figure p. 123.)

Ex.: "What is that called?"

Vis.: "The due guard of a Past Master."

Ex.: "Has that an allusion?"

Vis.: "It has, to the penalty of my obligation."

Ex.: "Give me another sign."

Visitor makes the sign of a Past Master. (See figure p. 123.)

Ex.: "What is that?"

Vis.: "The sign of a Past Master."

Ex.: "Has that an allusion?"

Vis.: "It has, to a portion of the penalty of my obligation."

Ex.: "Give me a grip."

Visitor gives the grip of a Past Master. (See pages 109-10.)

Ex.: "What is that?"

Vis.: "The grip of a Past Master."

Ex.: "Has it a name?"

Vis.: "It has."

Ex.: "Give it to me."

Vis.: "From a grip to a span—from a span to a grip."

Ex.: "What does it signify?"

Vis.: "Stone squarer."

The intelligent reader who has paid any attention at all to the initiatory ceremonies of Freemasonry can scarcely fail to notice that the pagan origin of Symbolic Masonry is indelibly stamped upon it, while he will as readily see that those we are now investigating—the Chapter degrees—are emphatically Jewish. In the Blue Lodge we have what is termed "The Ancient Mysteries;" in the Chapter we have Jewish things altogether. Solomon's Temple, Mount Lebanon, the quarries, the port of Joppa, the clay grounds between Succoth and Zarethan, the pillars of the porch, the dedication and destruction of the first Temple, the Babylonish captivity, the imitation Tabernacle, and the rebuilding of the Temple under Zerubbabel, after the captivity, are the leading incidents kept before us in the four Chapter degrees.

In Blue Lodge Masonry, on the contrary, we are ever reminded of the "Lesser Mysteries," by the Tyler's station, the hoodwink darkness, the cabletow, the meeting in "high places," and by the various initiatory rites—Preparation, Induction, Symbolic pilgrimage, Secrecy, Illumination, Intrusting, and Investiture; while in the Master Mason's degree we are introduced to the Greater Mysteries by the alleged conflict, death, burial, and raising of Hiram, constituting, as they do, the Egyptian legend of Osiris, or Baal, without even a single change. Turning, for example, to Sickles' "Freemason's Guide," (p. 196), we are emphatically assured that "Osiris and the Tyrian architect—Hiram Abiff—are one and the same, not a mortal individual, but an immortal principle." Hence, then, we are driven to the inevitable conclusion that the Masonic system is a horrible mixture of paganism and Jud while in the Royal Arch degree, as we shall see further along, the destruction of the first Temple is introduced, yet how strange it is, that the very religious philosophy and false worship which caused Jehovah to destroy His own temple, and banish into captivity His ancient people, are precisely the same philosophy and worship which modern Masons profess shall fit them for the glories of heaven. How extremely foolish it is, to imagine that what led to the utter destruction of God's earthly Temple, in the days of Zedekiah, will enable the Freemasons of to-day to "erect a spiritual temple, pure and spotless, and fit for the dwelling-place of Him who is the Author of purity." Surely there can be no greater delusion than to think that what God hated in the worship of the Israelites He will rejoice in, and accept, in the case of the Freemasons.

It seems to me that of all the stupid delusions in the world, Freemasonry is not only the greatest, but the most dangerous, and these Chapter degrees possibly establish that fact more clearly than any other part of the system. Making a loud display of "Trust in God," boasting of the Bible as a Masonic symbol, (though a symbol merely) and pretending to teach "piety, morality, science and self-discipline" (Webb's Monitor" by Morris, p. 17), it sets itself up as a saving religious philosophy, and hence the common declaration among the fraternity—"Freemasonry is a good enough religion for me." Throughout the entire system, however, the god of Masonry is Hiram Abiff, which is only the modern name of Baal, or Osiris of the Ancient Mysteries, while the disgusting and lascivious worship of the phallus is hidden beneath "the point within a circle."

At the same time Christ is rejected, the True God is unknown, the Holy Spirit is repudiated, and the Masonic system thus becomes the greatest enemy that Christianity has, and embodies within itself the true spirit of anti-Christ more thoroughly than any other form of religion that has ever appeared in the world; it is worse in that respect even than Romanism.

CHAPTER VI.

M. E. M. INTRODUCTION.

The origin and authorship of these preparatory degrees are entirely unknown. Tradition, the source of all Masonic knowledge, is altogether silent as to who manufactured the Most Excellent Master's degree, but A. T. C. Pierson, Past Grand Master of Minnesota, and a leading Masonic author, informs us that "the degree is purely an American invention," and adds further, that "it was first presented to the Masonic public in 1796-7." Like the third, fourth, and seventh degrees it is mainly dramatic, being at best but a clumsy caricature of the dedication of Solomon's Temple. It is needless to say, however, that King Solomon with all his wisdom knew nothing whatever of the ceremonies of this degree, though the majority of Royal Arch Masons believe the contrary. As proof of this, I may cite Dr. Albert Mackey, Past General Grand High Priest, who expressly declares that:

"The ceremonies of the degree as we now have them are not to be supposed to be the invention of King Solomon, or to have been known in his day. They are but a memorial subsequently established (at what later period we know not) of the events which occurred at the Temple."

To me it has always been a matter of astonishment, if not a source of much amusement, how men of intelligence, many of them lawyers, preachers, shrewd business men, and even professors and college presidents, will believe, or at least pretend to believe, the stupid nonsense and silly twaddle rehearsed for their edification in the various degrees of Freemasonry about Hiram Abiff, "the Keystone of the Arch," "the workmen from the quarries," the manner and hour of paying the Craft their wages, and a host of other legends, and worse than monkish fables, having not the slightest foundation in truth.

Take the case of Hiram Abiff as an example. Masonic tradition locates him at the Temple, "drawing designs upon his trestleboard," while the truth is, he was away from the Temple altogether, his particular line of work, that of brass moulder, confining him to "the clay grounds on the banks of the river Jordan, between Succoth and Zarethan," over fifty miles north-east of Jerusalem.

In 1 Kings vii: 13 we read: "And King Solomon sent, and fetched Hiram out of Tyre. He was the son of a widow woman (margin) of the tribe of Naphtali, and his father was a man of Tyre; a worker in brass; and he was filled with wisdom and understanding, and cunning to work all works in brass; and he came to King Solomon and wrought all his work." And the divine narrative goes on to enumerate the various things he made—the two brazen pillars, Boaz and Jachin, their chapiters and ornaments, the molten sea, the twelve oxen, the ten bases of brass, and their wheels of brass, the ten lavers of brass, "and the pots and the shovels and the basins;" the account closing with the emphatic declaration in verse 40: "So Hiram made an end of doing all the work that he made King Solomon for the house of the Lord." And, lest any doubt should arise about Hiram finishing all his work, the Second Book of Chronicles, iv: 11, especially states that "Hiram finished the work that he was to make for King Solomon for the house of the Lord."

Masonic tradition, however, declares that he did not finish the work, but was assassinated by three Fellow Crafts—Jubulum in particular—"before the Temple was completed," and so we find that through some strange delusion, or spiritual hypnotism, men, shrewd, intelligent and capable in all other matters, will accept the palpable falsehoods of mythology, and an absurd tradition, rather than the plain, consistent declaration of God's Word.

I speak of these matters here more particularly as in this degree I am now about to illustrate—that of MOST EXCELLENT MASTER—there are but two principal officers, namely, the Right Worshipful Master and the Senior Warden. The Junior Warden's chair is vacant because of the alleged murder of Hiram Abiff. The attention of the Craft is especially directed to this fact of the vacant chair in the South, and they need scarcely be reminded that the General Grand Chapter, in thus revising the secret work of this degree, has furnished a ritual at least consistent with itself, and in accordance with the traditions of the order. In other respects also the interior arrangement of a Most Excellent Masters' lodge is quite different from any of the preceding Chapter degrees.

In the previous degrees there is but one altar, in this degree there are two—the altar of obligation and the altar of incense—in this also there is a representation of the Ark and Mercy-seat, and an Arch set upon two pillars about five feet apart, and between five and six feet high. The Arch itself is in two sections, ready made, having a movable Keystone in the center, which is taken out when the Arch is first placed in position. These ready-made arches are furnished by all Masonic supply stores, and cost from $15.00 to $25.00, according to quality. In small country towns Chapter meetings are often held in rooms occupied by the Blue Lodge, and in that case the two pillars—Boaz and Jachin—inside the preparation room door, are used as the pillars of the Arch. (See diagram p. 161.)

DEGREE OF MOST EXCELLENT MASTER.

OPENING CEREMONIES.

As in the other preparatory degrees the aprons and jewels of office pertaining to the Royal Arch degree are worn in this, and the officers of the Chapter take rank as follows:

The High Priest as Right Worshipful Master; the King as Senior Warden; the Captain of the Host as Marshal, or Master of Ceremonies; the Principal Sojourner as Senior Deacon; The Royal Arch Captain as Junior Deacon, and the Treasurer, Secretary and Tyler occupy their usual places.

The Right Worshipful Master representing King Solomon ought to wear a crown and scarlet robe, as shown in the annexed figure, and as is some-

times worn during the Hiram Abiff tragedy in the
Master Mason's degree. Being thus clothed, and
the officers and members having resumed their
aprons and jewels of office, the

R. W. M. gives one rap, the brethren are seated
and the Junior Deacon closes the ante-room door.

R. W. M. (one rap): "Brother Senior Warden,
are you satisfied that all present are Most Excellent
Masters?"

S. W. Looking around the room and noticing
no stranger present, reports:

S. W. "All present are Most Excellent Masters, Right Worshipful."

Should he not know all the brethren however, he replies:

S. W.: "I shall ascertain by the proper officers and report. Brothers Senior and Junior Deacons approach the West and give me the pass of a Most Excellent Master."

The Senior and Junior Deacons meet at the west side of the altar, as in the preceding degrees, repair to the Senior Warden's station and whisper the pass-word, Rabboni, in his ear.

S. W.: "You will now (he continues) collect the pass from the brethren and report again to the West."

The Senior and Junior Deacons pass along their respective sides of the lodge, the Senior on the North side, and the Junior on the South, and as they approach, each brother rises to his feet and whispers in their ears the pass-word, Rabboni. Having thus collected the pass, they return and report:

S. D.: "All are Most Excellent Masters in the North, brother Senior Warden."

J. D.: "All are Most Excellent Masters in the South."

They then retire to their seats, and the Senior Warden reports:

S. W.: "All present are Most Excellent Masters, Right Worshipful;" and takes his seat."

R. W. M. (one rap): "Brother Junior Deacon, what is the first great care of assembled Masons?"

J. D. (standing): "To see that the lodge is duly tyled, Right Worshipful."

R. W. M.: "Perform that duty. Inform the Tyler that I am about to open a lodge of Most Excellent Masters; direct him to take due notice and tyle accordingly."

The Junior Deacon opens the door and whispers to the Tyler that the lodge is about being opened; closes it again and gives six knocks, by two's, on the door—each double knock quite rapidly, and then a slight pause.

The Tyler knocks in a similar manner on the outside.

J. D.: "Right Worshipful Master, that duty is performed, and the lodge is duly tyled."

R. W. M.: "How is it tyled, brother Junior Deacon?"

J. D.: "By a worthy brother without. armed with the proper implement of his office."

R. W. M.: "His duty there?"

J. D.: "To keep off all cowans and eavesdroppers, and admit none but such as are duly qualified and have permission from the Right Worshipful Master."

R. W. M. Gives one rap and J. D. is seated.

R. W. M.: "Brother Senior Warden, are you a Most Excellent Master?"

S. W. (rising): "I am; try me."

R. W. M.: "How will you be tried?"

S. W.: "By the Keystone."

R. W. M.: "Why by the Keystone?"

S. W.: "Because at the completion and dedication of King Solomon's Temple, and the placing of the Keystone, this degree was founded."

R. W. M.: "Where were you made a Most Excellent Master?"

S. W.: "In a legally constituted and duly opened lodge of Most Excellent Masters."

R. W. M.: "How many, compose a lodge of Most Excellent Masters?"

S. W.: "Two, or more."

R. W. M.: "When composed of only two, who are they?"

S. W.: "The Right. Worshipful Master and Senior Warden."

R. W. M.: "Why is there no Junior Warden?"

S. W.: "Because just before the completion of King Solomon's Temple. our Grand Master Hiram Abiff was assassinated, as we have an account in a preceding degree. , At the dedication thereof, as no one had been appointed to fill his place, his station was vacant and his light extinguished."

R. W. M.: "The Senior Warden's station?"

S. W.: "In the West, Right Worshipful."

R. W. M.: "Why in the West, and your duty there?"

S. W.: "As the sun is in the West at the close of the day, so is the Senior Warden in the West. to assist the Right Worshipful Master in opening and closing the lodge, to pay the Craft their wages, if aught be due, and see that none go away dissatisfied: harmony being the strength and support of all well-governed institutions."

R. W. M.: "The Right Worshipful Master's station?"

S. W.: "In the East."

R. W. M.: "Why in the East, and his duty there?"

S. W.: "As the sun rises. in the East to open and govern the day, so is the Right Worshipful Master in the East to open and govern the lodge, set the Craft to work, and give them proper instruction for their labor."

R. W. M. (Rising, gives three raps, calling up the entire lodge): "Brother Marshal, you will assemble the brethren around the altar, and see that they are in due form for our devotions."

Marshal: "Brethren, you will assemble around the altar and be in due. form for our devotions. Kneel on your right knee, forming a chain with your right hands over the left arm."

The brethren assemble around the altar in a circle, as directed, standing close together, and leaving an opening at the East and West sides of the circle for the Right Worshipful Master and Senior Warden respectively. They then kneel on the right knee and cross the right arm over the left, taking hold with the right hand of the left hand of the brother on the left, and with the left hand the right hand of the brother on. the right, thus forming a chain around the altar, with an open space, East and West, for the two principal officers. All being "in due form," the Marshal announces:

Marshal: "Right Worshipful Master, the brethren are in due form for our devotions, and await your presence." He then takes his own place.

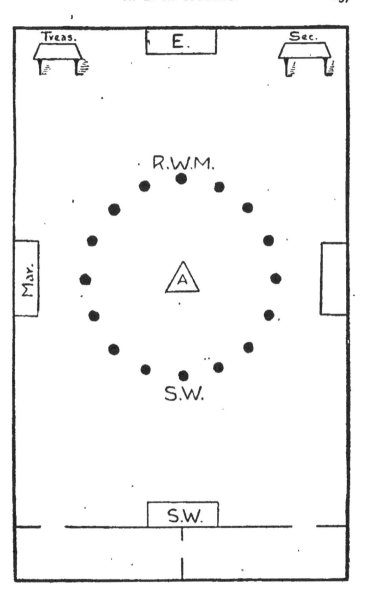

R. W. M. and S. W. then descend and fill up the spaces left open at the East and West sides of the altar, and complete the chain. (See p. 137.)

All now repeat the Lord's prayer together, following the lead of the R. W. M.; then they balance six times with their arms as the Right Worshipful Master counts one, two, three. That is, retaining the position of the arms they raise them up. then down, up again, and down; the R. W. M timing each movement thus:—"one, two, three; one, two, three;" so that the balancing with the arms will be as follows: up, down, up; down, up, down. They then disengage their hands, all rise; and the Right Worshipful Master and Senior Warden return to their stations; the others remain standing.

R. W. M.: "Brother Senior Warden, it is my order that a lodge of Most Excellent. Masters be now opened for the dispatch of business under the usual Masonic restrictions. This order you will communicate to the brethren for their government."

S. W.: "Brethren, it is the order of the Right Worshipful Master that a lodge of Most Excellent Masters be now opened for the dispatch of business under the usual Masonic restrictions; of this you will take due notice and govern yourselves accordingly. Look to the East."

In concert, with the Right Worshipful Master, they now make all the signs, from the Entered Apprentice degree to that of Most Excellent Master inclusive. (For the signs of preceding degrees see pages 12, 13 and 103.)

The Right Worshipful Master gives two quick raps; the Senior Warden the same. (** **)

The R. W. M. again gives two raps; the S. W. likewise.

Again the R. W. M. raps twice, and the S. W. twice; each of these officers thus giving six raps, corresponding to the number of this degree in American Freemasonry.

The Right Worshipful Master, or Chaplain, then reads from the Manual of the Chapter the 24th Psalm:

"The earth is the Lord's and the fulness thereof: the world, and they that dwell therein. For he hath founded it upon the seas, and established it upon the floods. Who shall ascend into the hill of the Lord? or who shall stand in his holy place? He that hath clean hands and a pure heart; who hath not lifted up his soul unto vanity, nor sworn deceitfully. He shall receive the blessing from the Lord, and righteousness from the God of his salvation. This is the generation of them that seek him, that seek thy face, O Jacob: Lift up your heads, O ye gates; and be ye lifted up, ye everlasting doors, and the King of Glory shall come in. Who is this King of Glory? The Lord, strong and mighty; the Lord, mighty in battle. Lift up your heads, O ye gates; even lift them up, ye everlasting doors, and the King of Glory shall come in. Who is this King of Glory? The Lord of Hosts, he is the King of Glory."

R. W. M.: (still standing): "I now declare this lodge of Most Excellent Masters, erected to God and dedicated to the memory of King Solomon, open in due form. Brother Junior Deacon, inform the Tyler. Brother Senior Deacon, display the

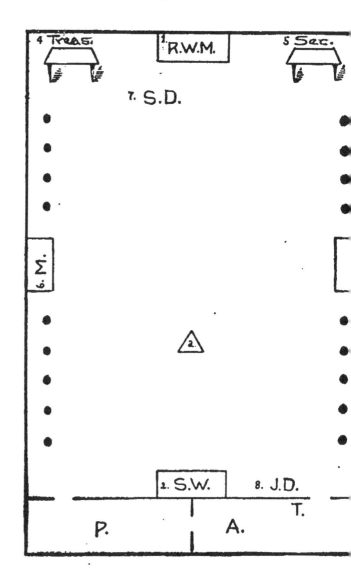

great lights." He then gives one rap and all are seated as represented in diagram p. 140, as follows:

1. Right Worshipful Master. 2. Senior Warden. 3 Vacant. 4. Treasurer. 5. Secretary. 6. Marshal. 7. Senior Deacon. 8. Junior Deacon. P. Preparation room. A. Ante-room. T. Tyler. ● Members. a. Altar of obligation.

J. D. gives six raps on the ante-room door s at opening—three double knocks with a pause between every two. (**. ** **) The Tyler answers in the

Due Guard.

same manner, and opens the door. The J. D. whispers that "the lodge is now open," closes the door again, and gives six raps as before, which are again answered by the Tyler.

J. D. (facing the East and making the due guard and sign): "The Tyler is informed, Right Worshipful;" and takes his seat.

The due guard of a Most Excellent Master is made by placing both hands with the fingers half-closed at the middle of the breast, the ends of the fingers touching the vest, as in figure p. 141.

Then draw the hands swiftly apart and drop them by your side. This is the penal sign, and alludes to the penalty of the obligation in this degree, namely, to have the breast torn open and the heart and vitals taken from thence, etc.

These signs are always given together, and are used as a salute to the Right Worshipful Master when entering or retiring from a lodge of Most Excellent Masters, or when addressing the East.

CHAPTER VII.

MOST EXCELLENT MASTER.

FIRST SECTION.

SECRET WORK.

R. W. M.: "Brethren, this lodge of Most Excellent Masters has been called and opened for the purpose of receiving and acknowledging as a Most Excellent Master, brother James Hunt, who has been duly elected to receive this degree, and if there be no objections we will now proceed with the ceremonies of reception." Pausing for a second or two, he continues: "There being none, it is so ordered."

R. W. M.: "Brother Junior Deacon, you will see that the Candidate is prepared and presented."

The Secretary having already collected the initiation fee, the Junior Deacon retires to the Preparation room, and prepares the Candidate by investing him with an apron—a Past Master's apron being the proper one to be used. He then ties the cable-tow six times about his body,* in which condition he is led to the door, upon which the Junior Deacon gives six knocks in the usual manner, by two's. (** ** **.)

---

* The cable-tow used in these preparatory degrees are from eight to ten yards long, and the coils about the body of the preacher correspond to the number of each degree to signify that "as he advances in Masonry his obligations become more and more binding."

S. D.: "Right Worshipful Master, there is an alarm at the door of the Preparation room."

R. W. M.: "Attend to that alarm, brother Senior Deacon."

The Senior Deacon goes to the door, upon which he gives the usual six raps, opens the door, and says:

S. D.: "Whʊ ːomes here?"

J. D.: "Brother James Hunt, who has been regularly initiated, passed and raised to the sublime degree of a Master Mason, advanced to the honorary degree of a Mark Master, has been inducted into the Oriental Chair of King Solomon, and now seeks further promotion in Masonry by being received and acknowledged a Most Excellent Master."

S. D.: "Brother James Hunt, is it of your own free will and accord that you make this request?"

Candidate: "It is."

S. D.: "Brother Junior Deacon, is he duly and truly prepared?"

J. D.: "He is."

S. D.: "Is he worthy and well qualified?"

J. D.: "He is."

S. D. "Has he made suitable proficiency in the preceding degrees?"

J. D.: "He has."

S. D.: "By what further right or benefit does he expect to gain admission?"

J. D.: "By the benefit of the pass."

S. D.: "Has he the pass?"

J. D.: "He has it not, but I have it for him."

S. D.: "Advance and give me the pass;" or "Advance and give it."

The Junior Deacon steps up close to the Senior Deacon and whispers in his ear the word, Rabboni.*

S. D.: "Let him wait until the Right Worship-ul Master is informed of his request, and his answer returned."

The Senior Deacon closes the door, returns to the west of the altar, makes the due guard and sign --saluting the Right Worshipful Master—and reports.

R. W. M.: "Brother Senior Deacon, what was the cause of that alarm?".

S. D.: "Right Worshipful Master, there is without brother James Hunt, who has been regularly initiated, passed, and raised to the sublime degree of a Master Mason, advanced to the honorary degree of a Mark Master, has been inducted into the Oriental Chair of King Solomon, and now seeks further promotion in Masonry by being received and acknowledged a Most Excellent Master."

R. W. M.: "Is it of his own free will and accord that he makes this request?"

S. D.: "It is."

R. W. M.: "Is he duly and truly prepared, worthy and well qualified?"

S. D: "He is."

R. W. M.: "Has he made a suitable proficiency in the preceding degrees?"

---

* Masonry blackballs the Lord Jesus personally from its lodges and Chapters and uses in mockery as a pass word, one of the most endearing titles by which He was addressed—Rabboni—beloved Lord, thus as it were betraying Him with a kiss, and the preacher consents.

S. D.: "He has."

R. W. M.: "Who vouches for this?"

S. D.: "A brother."

R. W. M.: "By what further right or benefit does he expect to gain admission?"

S. D.: "By the benefit of the pass."

R. W. M.: "Has he the pass?"

S. D.: "He has it not, but his conductor gave it for him."

R. W. M.: "Give me the pass."

S. D. (with due guard and sign): "Rabboni."

R. W. M.: "The pass is right; you will admit the Candidate and receive him in due form."

The Senior Deacon salutes the Right Worshipful Master, as before, with the due guard and sign, returns to the door, opens it up wide, and says:

S. D.: "It is the order of the Right Worshipful Master that the Candidate enter and be received in due form."

The Junior Deacon, taking the Candidate by the right arm, conducts him into the lodge room, and about three or four feet inside the door, where he is met by the Senior Deacon, who halts him and places the small end of the Keystone against his breast, and says: (See figure p. 147.)

S. D.: "Brother Hunt, I receive you into this Most Excellent Master's lodge upon the Keystone, because at the completion and dedication of the Temple the stone which the builders had rejected became the head stone of the corner."

He then hands the Keystone to the Junior Deacon, takes the Candidate with his left hand by the right arm, and conducts him six times about the lodge room, keeping the altar on the right, the Junior Deacon walking behind until he reaches his place in the lodge when he takes his seat.

As they pass the West for the first time the Senior Warden gives one rap with his gavel. Arriv-

ing at the East the Right Worshipful Master gives one rap. The second time around the Senior Warden gives two quick raps, and, passing the chair, the Master gives two. The third time going around, the S. W. gives two quick raps, and then one; the R. W. M., as they pass him, does the same. The next time,

the S. W. gives four raps—two and two, the Master
the same. The fifth time around, the S. W. gives
five raps—two—two—one; the Right Worshipful
Master likewise, and at the sixth revolution, the S.
W. gives six raps—two— two—two, and the Master
gives the same number and in the same way. (**
** **.) Let it be noticed that the Senior Warden
leads off in these raps, and that the raps are not
given until the Senior Deacon and Candidate have
arrived in front of the station, either West or East.

During the pilgrimage of the Senior Deacon
and Candidate about the lodge room, the Right
Worshipful Master reads from Psalm cxxii in the
following order, namely:

First time around: "I was glad when they said
unto me, Let us go into the house of the Lord."

Second: "Our feet shall stand within thy gates,
O Jerusalem. Jerusalem is builded as a city that is
compact together."

Third: "Whither the tribes go up, the tribes of
the Lord, unto the testimony of Israel, to give
thanks unto the name of the Lord."

Fourth: "For there are set thrones of judgment, the thrones of the house of David."

Fifth: "Pray for the peace of Jerusalem; they
shall prosper that love thee. Peace be within thy
walls, and prosperity within thy palaces."

Sixth: "For my brethren and companions'
sakes I will now say, Peace be within thee. Because
of the house of the Lord our God I will seek thy
good."—R. A. Standard, p. 51.

The reading of each verse is so timed that it is finished when the Candidate has made the circuit of the room, and the entire Psalm is concluded when the Senior Deacon and Candidate have made six complete circuits.

The number of circuits in each degree denotes the number of that degree. Thus, in the first degree they go round once; in the second degree twice, and in the third three times, etc.

The Senior Deacon and Candidate, having thus made six complete revolutions, arrive at the Junior Warden's station where the

S. D. gives six knocks, but there is no response, the chair being empty. A moment's silence ensues, and then the Senior Warden replies:

S. W.: "Alas, poor Hiram! unexpectedly, in the midst of life, duty and usefulness, he was overtaken by death; suddenly his sun of life went down at noon, and his station is vacant. Let us remember his virtues, and imitate his worthy example; may we be reminded of the shortness of life, and the uncertainty of its continuance, remembering that soon, when our brethren shall assemble to labor, our stations will also be vacant forever. May we all at last fill our appropriate stations in heaven." *

The Senior Deacon next leads the Candidate to the Senior Warden in the West.

S. D. gives six knocks as before—two, two, two. (** ** **.)

---

* "Let this mind be in you which was also in Christ Jesus," (Phil ii: 5) is the rule set before the Christian; "let us imitate the worthy ex ample of Hiram Abiff " is that which is set before the Mason. Always Hiram against Christ.

S. W.: "Who comes here?"

S. D.: "Brother James Hunt, who has been regularly initiated, passed, and raised to the sublime degree of Master Mason, advanced to the honorary degree of Mark Master, has been inducted into the Oriental Chair of King Solomon and now seeks further promotion in Masonry by being received and acknowledged a Most Excellent Master."

S. W.: "Brother Hunt, is it of your own free will and accord that you make this request?"

Can.: "It is."

S. W.: "Brother Senior Deacon, is the Candidate duly and truly prepared?"

S. D.: "He is."

S. W.: "Is he worthy and well qualified?"

S. D : "He is."

S. W.: "Has he made suitable proficiency in the preceding degrees?"

S. D.: "He has."

S. W.: "By what further right or benefit does he expect to obtain this favor?"

S. D.: "By the benefit of the pass."

S. W.: "Has he the pass?"

S. D.: "He has it not, but I have it for him."

S. W.: "Give me the pass."

The Senior Deacon whispers to the Senior Warden the word—Rabboni—as before.

S. W.: "You will conduct the Candidate to the Right Worshipful Master in the East for final examination."

The Senior Deacon conducts the Candidate to the East, and gives six knocks in front of the Master's chair.

R. W. M.: "Who comes here?"

S. D.: "Brother James Hunt, who has been regularly initiated, passed, and raised to the sublime degree of a Master Mason, advanced to the honorary degree of a Mark Master, has been inducted into the Oriental Chair of King Solomon, and now seeks further promotion in Masonry by being received and acknowledged a Most Excellent Master."

R. W. M.: "Brother Hunt, is it of your own free will and accord that you make this request?"

Can.: "It is."

R. W. M.: "Brother Senior Deacon. is the Candidate duly and truly prepared, worthy and well qualified?"

S. D.: "He is."

R. W. M.: "Has he made suitable proficiency in the preceding degrees?"

S. D.: "He has."

R. W. M.: "By what further right or benefit does he expect to obtain this favor?"

S. D.: "By the benefit of the pass."

R. W. M.: "Has he the pass?"

S. D.: "He has it not, but I have it for him."

R. W. M.: "Give me the pass."

S. D. whispers in the Right Worshipful Master's ear the word, Rabboni.

R. W. M.: "The pass is right. You will reconduct the Candidate to the West and place him in charge of the Senior Warden, who will teach him how to approach the East in a proper manner."

The Senior Deacon re-conducts the Candidate to the west side of the altar, and about eight feet from it, facing the Senior Warden's station, and giving one knock on the floor, says:

S. D.: "Brother Senior Warden, it is the order of the Right Worshipful Master that you teach the Candidate how to approach the East in a proper manner."

S. W.: "Brother Senior Deacon, face the Candidate to the East."

The Senior Deacon faces the Candidate about.

S. W.: "Brother Hunt, you will approach the East with six steps—first, as an Entered Apprentice;" he takes that step. "Second, as a Fellow Craft;" he takes the Fellow Craft step. "Third, as a Master Mason;" he steps off as a Master Mason. "Fourth, as a Mark Master;" he takes the Mark Master's step. "Fifth, as a Past Master." This is done. (See p. 108.) The Senior Deacon sees to it that he makes every step as ordered. The Senior Warden then continues:

S. W.: "You will now advance one step with your right foot, bring the heel of the left to the heel of the right, the toes turned out equally, the feet forming a right angle."

The Senior Deacon assists in placing the Candidate's feet as required.

S. W.: "Right Worshipful Master, the Candidate is before the altar."

R. W. M.: "My brother, for the sixth time you are before the altar of Freemasonry, and before you can proceed further it is necessary that you take upon yourself a solemn obligation pertaining to this degree. which contains nothing that will conflict with the duty you owe to God, your country, your family, or yourself. With this assurance on my part, do you wish to proceed?"

Can.: "I do."

R. W. M.: "Brother Senior Deacon, you will place the Candidate at the altar in due form."

The Senior Deacon leads the Candidate to the altar, where he kneels on both knees, both hands resting on the Holy Bible, Square and Compass, and his body erect.

S. D.: "The Candidate is in due form, Right Worshipful."

R. W. M. gives three raps. All the brethren arise and form two parallel lines from the altar to the East, facing each other, as in the preceding degrees. The Senior Deacon stands a little behind and towards the Candidate's right. The Right Worshipful Master removing his crown, approaches the east side of the altar, facing the Candidate and says: (See diagram, p. 154.)

R. W. M.: "You will say I, pronounce your name, and say after me:"

OATH OF MOST EXCELLENT MASTER.

Can.: "I, James Hunt"—and repeats after the R. W. M.—

R. W. M.: "of my own free will and accord, in the presence of Almighty God and this Most Excellent Masters' lodge, erected to God and dedicated

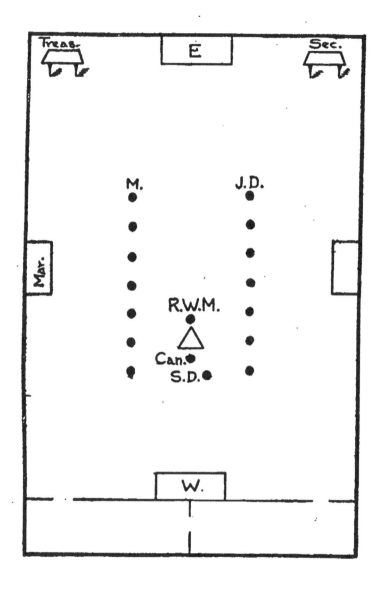

to the memory of King Solomon, do hereby and hereon sincerely promise and solemnly swear that I will not reveal the secrets of this degree to a brother of a preceding degree, nor to any person in the world, except it be within a legally constituted and duly opened Most Excellent Masters' lodge, or to a brother of this degree whom I shall have found to be such by due trial, strict examination, or legal information."

"I do furthermore promise and swear that I will answer and obey all due signs and summons sent to and received by me from a Most Excellent Masters' lodge, or given me by a brother of this degree, if within the length of my cable-tow."

"I do furthermore promise and swear that I will help, aid and assist all worthy and distressed Most Excellent Masters, their wives, widows and orphans, wherever I may find them, so far as their necessities may require and my ability permit without injury to myself or those having prior claims upon me."

"I do furthermore promise and swear that I will dispense true Masonic light and knowledge to my less informed brethren to the best of my ability."

"I do furthermore promise and swear that I will not derogate from the character I am about to assume, it being that of a Most Excellent Master."

"All of this I sincerely promise and swear, without the least hesitation, equivocation or mental reservation, binding myself under no less a penalty than that of having my breast torn open, my vitals taken from thence and thrown upon a dung-hill to rot, should I violate this my solemn obligation as a

Most Excellent Master.   So help me God, and keep me steadfast."

R. W. M.: "In token of your sincerity, you will detach your hands and kiss the Holy Bible now open before you."   Candidate kisses the open Bible.*

R. W. M.: "The brother being now bound to us by a tie that cannot be broken, brother Senior Deacon, you will release him from the cable-tow." (The cable-tow is removed.)

The Right Worshipful Master now steps back towards his chair, resumes his crown, and, standing at the end of the parallel lines of brethren, facing the Candidate, says:

R. W. M. (addressing Candidate): "You now behold me approaching you from the East with the step" (takes the step,) "due guard" (makes the due guard,) "and penal sign" (gives the sign,) "of a Most Excellent Master; extending to you my right hand in token of continued friendship, brotherly love and confidence," (he takes the Candidate by the right hand and raises him.) "Raising you from a right angle to a perpendicular." Stepping back again a few paces, he continues:

R. W. M.: "This is the step of a Most Excellent Master." Step off one step with the right foot, and bring the heel of the left to the heel of the right, toes equally extended and feet forming a right angle.

"This is the due guard and sign of a Most Excellent Master;" makes these signs as already ex-

* Can any one show what there is in Freemasonry to warrant the taking of such terrible oaths with their blood-curdling death penalties in order to conceal it? The Chinese boxers might swear such oaths as these, but that an American citizen, and especially a preacher should do so, is almost incredible.

plained. (p. 141.) "They allude to the penalty of your obligation wherein you have sworn to have your breast torn open, your vitals taken from thence," etc. "You will always make this due guard and sign as a salute to the Right Worshipful Master upon entering or retiring from a lodge of Masons working on this degree."

"With the assistance of the Senior Deacon, I will now instruct you in the grip and word of a Most Excellent Master. Take me as I take you."

The Right Worshipful Master takes the Candidate by the right hand and presses his thumb on the third knuckle-joint as in the annexed figure. The

Senior Deacon places the Candidate's thumb on the Right Worshipful Master's hand in the same manner.

R. W. M.: "Brother Senior Deacon, you will respond. What is that?"

S. D.: "The grip of a Most Excellent Master."

R. W. M.: "Has it a name?"

S. D.: "It has."

R. W. M.: "Will you give it to me?"

S. D.: "I did not so receive it, and cannot so impart it."

R. W. M.: "How will you dispose of it?"

S. D.: "I will syllable it with you."

R. W. M.: "Syllable it, and begin."

S. D.: "Ra."

R. W. M.: "Bbo.'

S. D.: "Ni."

R. W. M.: "Rab."

S. D.: "bo."

R. W. M.: "ni."

S. D.: "Rabboni."

R. W. M. (to Candidate): "This is the grip of a Most Excellent Master, and its name is Rabboni. It is otherwise called the covering grip, because as this covers the grips of preceding degrees, so ought we, as Most Excellent Masters considering that man in his best estate is subject to frailty and errors. endeavor to cover his faults with the broad mantle of charity and brotherly love."

The Right Worshipful Master then returns to the East; the Senior Warden is at his station in the West; the brethren are standing in two parallel lines, and the Senior Deacon and Candidate are still standing to the west of the altar. And thus the first section of the degree is ended.

In preceding degrees, the working tools were presented to the Candidate at this stage of the ceremonies, but in this degree there are no working tools, the Temple being supposed to have been finished, and hence working tools are not required. The Right Worshipful Master therefore continues at once.

SECOND SECTION.

R. W. M.: "Brother Senior Warden, this is the day and the hour set apart for celebrating the Cape-stohe, and seating the Ark of the Covenant; you will assemble the Craft for the purpose of proceeding in these solemn ceremonies."

S. W.: "Brother Marshal, you will assemble the Craft for the purpose of proceeding in the solemn ceremonies of celebrating the Cape-stone and seating the Ark of the Covenant."

Marshal: "Brethren, you will form in procession on the north side of the lodge, double file, facing the East."

The brethren stand as directed, the Right Worshipful Master and Senior Warden at the head of the procession, and the Senior Deacon and Candidate bringing up the rear. The procession moves slowly around the room, and sometimes pass into the anteroom; the Senior Deacon and Candidate as they come to it procure the Keystone and bear it between them while all join in singing the following ode:

"MOST EXCELLENT MASTER'S SONG.

"All hail to the morning that bids us rejoice;
  The Temple's completed, exalt high each voice;
  The cape-stone is finished, our labor is o'er;
  The sound of the gavel shall hail us no more.

"To the power Almighty, who ever has guided
  The tribes of old Israel, exalting their fame;
To Him who hath governed our hearts undivided,
  Let's send forth our voices to praise His great name.

"Companions assemble on this joyful day
  (The occasion is glorious) the keystone to lay;
Fulfilled is the promise, by the ANCIENT OF DAYS,
  To bring forth the cape-stone with shouting and praise."

The marching and singing are so timed that the brethren have made three circuits of the room by the time the third stanza is finished, and are standing in two rows from West to East a little to the north side of the altar. At the end of the third stanza—"with shouting and praise"—the procession halts, the brethren open ranks, and the Senior Deacon and Candidate pass through, bearing between them the imitation keystone. The open ranks of the brethren are close to and almost directly in a line with the Arch already explained on page 131, and

as the Senior Deacon and Candidate arrive at the end of the open column, the Right Worshipful Master takes the Keystone and assisted by the Senior Warden places it in the center of the arch— the part purposely left to receive it—and strikes upon it six blows with his gavel. (See diagram p. 161.)

This ceremony being completed, the Right Worshipful Master, turning to look at the appearance of the Keystone in its place, raises both hands, and

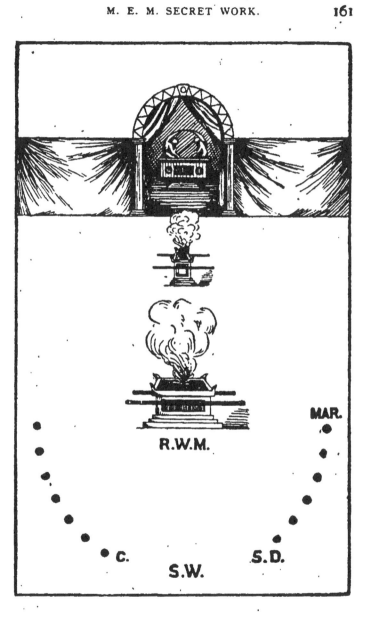

throws them forward a little, as if expressing admiration or astonishment; all the brethren do the same, in concert with the R. W. M., and all exclaim Rabooni! (See figure.) The Senior Deacon and Candidate go back to the west of the column again.

R. W. M.: "The Temple is completed, and nothing now remains to be done but to seat the Ark of the Covenant within the Sanctuary."

The procession is again formed as before, the Master and Warden being at the head, the Senior Deacon and Candidate bringing up the rear, and bearing the imitation Ark assisted by two other brethren. They march three times around the lodge room, the third time the column extending from West to East on the north side of the altar as when carrying the Keystone. The Marshal opens the

ranks, passes through towards the West where the
Ark is being borne, and says:

Marshal: "Brethren advance with the Ark."

The four brethren bearing the Ark—the Candidate being one of them—pass through the lines and
under the Arch in which the Keystone was placed,
and deposit the Ark on a pedestal inside, toward the
East as accurately represented in the diagram. The
Ark is now supposed to be seated in the Holy of

Holies. The place where the altar of incense stands
represents the Sanctuary, while the portion of the
lodge room containing the altar of obligation is supposed to be the Court, but of these things the majority of the Craft know little or nothing.

The procession, in the meantime, breaks up and
the brethren form a semi-circle around the west side
of the altar facing the East.

The diagram on page 161 furnishes a very good
illustration of the Arch, the Ark, and the relative

position of the Master and brethren at this stage of the ceremonies. ("R. A Standard" p. 46.) The singing of the remaining stanzas of the ode is now resumed.

"There is no more occasion for level or plumb-line,
    For trowel or gavel, for compass or square;
    Our works are completed, the Ark safely seated,
    And we shall be greeted as workmen most rare."

During the singing of this verse the brethren take off their aprons, but they don't throw them on the altar, nor hang them on the arch, for they are not near the arch; they simply lay them down on the chairs most convenient to them.

"Now those that are worthy, our toils who have shared,
    And proved themselves faithful, shall meet their reward;
    Their virtue and knowledge, industry and skill,
    Have our approbation, have gained our good will."

The Senior Deacon and Candidate stand opposite the Senior Warden's station, (See diagram) and as the brethren begin to sing this verse they move on slowly once more, single file, the Right Worshipful Master first, the singing being so timed that the R. W. M. reaches the Candidate when the next verse is begun:

"We accept and receive them, Most Excellent Masters,
    Invested with honors and power to preside
    Among worthy Craftsmen, wherever assembled,
    The knowledge of Masons to spread far and wide."

The Right Worshipful Master and each of the brethren, as they pass in front of the Candidate during the singing of this verse, take his hand, giving him the Most Excellent Master's grip, and keep on marching slowly. When the last man passes the

Candidate, both he and the Senior Deacon fall in behind, the singing being continued:

"Almighty Jehovah, descend now and fill
 This lodge with thy glory, our hearts with good-will;
 Preside at our gatherings, assist us to find
 True pleasure in teaching good-will to mankind." *

About the time this stanza is finished, the head of the procession ought to have reached the East, and as they begin the next verse the Right Worshipful Master turns in toward the altar, the brethren moving on, and by the time the verse is finished, they stand in a sort of semi-circle about the altar, as shown in the diagram on p. 161, the R. W. M. being at the west side of the altar. .

"Thy wisdom inspired the great institution,
 Thy strength shall support it till nature expire;
 And when the creation shall fall into ruin,
 Its beauty shall rise through the midst of the fire."

All are now standing, and the R. W. M. reads from the Monitor:

R. W. M.: "The Lord hath said that He would dwell in the thick darkness, but I have built an house of habitation for Thee and a place for Thy dwelling forever." (2 Chron. vi: 1, 2.)

R. W. M.: "And he said, Blessed be the Lord God of Israel, who hath with his hands fulfilled that which he spake with his mouth to my father David, saying,

"Since the day that I brought forth my people out of the land of Egypt, I chose no city among all

---

* It is impossible to conceive of a greater mockery than this when the spiritual condition of the majority of those present is considered, and the preacher who joins in it is no better than the rest. Assuredly if Masonry be true he ought to quit preaching.

the tribes of Israel to build an house in, that my name might be there; neither chose I any man to be a ruler over my people Israel:

"But I have chosen Jerusalem, that my name might be there; and have chosen David to be over my people Israel.

"Now it was in the heart of David my father to build an house for the name of the Lord God of Israel.

"But the Lord said to David my father, Forasmuch as it was in thine heart to build an house for my name, thou didst well in that it was in thine heart:

"Notwithstanding, thou shalt not build the house; but thy son which shall come forth out of thy loins, he shall build the house for my name.

"The Lord therefore hath performed his word that he hath spoken; for I am risen up in the room of David my father, and am set on the throne of Israel, as the Lord promised, and have built the house for the name of the Lord God of Israel:

"And in it have I put the ark, wherein is the covenant of the Lord, that he made with the children of Israel."

This portion concluded, all the brethren except the Master kneel, and he again reads from the Monitor:

R. W. M.: "O Lord God of Israel, there is no god like thee in the heaven, nor in the earth; which keepest covenant, and showest mercy unto thy servants that walk before thee with all their hearts:

"Thou which hast kept with thy servant David my father that which thou hast promised him: and spakest with thy mouth, and hast fulfilled it with thine hand, as it is this day.

"Now therefore, O Lord God of Israel, keep with thy servant David my father that which thou hast promised him, saying, There shall not fail thee a man in my sight to sit upon the throne of Israel; yet so that thy children take heed to their way to walk in my law, as thou hast walked before me.

"Now then, O Lord God of Israel, let thy word be verified, which thou hast spoken unto thy servant David.

"But will God in very deed dwell with men on the earth! Behold, heaven and the heaven of heavens cannot contain thee; how much less this house which I have built!

"Have respect, therefore, to the prayer of thy servant, and to his supplication, O Lord my God, to hearken unto the cry and the prayer which thy servant prayeth before thee:"

"That thine eyes may be open upon this house day and night, upon the place whereof thou hast said that thou wouldest put thy name there; to hearken unto the prayer which thy servant prayeth toward this place.

"Hearken, therefore, unto the supplications of thy servant, and of thy people Israel, which they shall make toward this place: hear thou from thy dwelling-place, even from heaven: and when thou hearest, forgive.

"Now, my God, let, I beseech thee, thine eyes be open, and let thine ears be attent unto the prayer that is made in this place.

"Now therefore arise, O Lord God, into thy resting place, thou, and the ark of thy strength: let thy priests, O Lord God, be clothed with salvation, and let thy saints rejoice in goodness.

"O Lord God, turn not away the face of thine anointed: remember the mercies of David thy servant."—R. A. Monitor, p. 117. (Sheville & Gould.)

As these last words are read, fire "comes down from above" and ignites the pot of incense upon the

altar; the kneeling brethren bow their faces to the floor and repeat three times: "For he is good; for his mercy endureth forever."

It will be understood, of course, that this fire from "above," mentioned here, is produced by a chemical, or as the ingenuity of the Master may suggest.

The kneeling and bowed down brethren, having made those three exclamations at the flash of fire in the little pot of incense, all rise to their feet. The Right Worshipful Master retires to his seat, gives one rap, and all are seated—the Senior Deacon and Candidate stand at the west side of the altar.

R. W. M.: "Brother Senior Deacon, you will conduct the Candidate to the East."

The Candidate is led to a seat in front of the Right Worshipful Master's chair, who then reads or rehearses to the Candidate as much of the historical summary of the Most Excellent Master's degree contained in the Monitor as he may think best. Usually, he reads or repeats the following:

R. W. M.: "My brother, in the course of the ceremonies you have received an additional sign to that given you at the altar; it is this:" Makes the sign of admiration. (See figure p. 162.)

"It is called the sign of admiration, and alludes to the wonder and admiration expressed by those of our ancient brethren who were permitted to view the interior of that magnificent edifice which King Solomon had erected and was about to dedicate to the service of the Supreme Being.

"The ceremonies of this degree are intended to represent those of the completion and dedication of King Solomon's Temple. You have now arrived at a period in Masonry when the labor is over. The Keystone has been placed in the principal arch, the Temple finished, and the ark, which had been so long without a permanent resting place, is at last safely seated.

"We have imitated our ancient brethren in assembling on that occasion, repairing to the place designated, and participating in those solemn ceremonies. We have imitated them in gathering around the altar, engaging in prayer, and have witnessed a representation of the fire coming down from heaven, consuming the burnt-offering and the sacrifices. We have also imitated their astonishment on beholding it, by falling down upon the ground and exclaiming: 'For He is good; for His mercy endureth forever!'

"A perusal of the books of Chronicles and Kings will give you a minute description of the Temple and of the ceremonies here intended to be represented.

"You will there find that the foundations of the Temple were laid by King Solomon in the year of the world 2992, and the building was finished in the year 3000. About seven years and six months were consumed in its erection.

"It was dedicated in the year 3001, with the most imposing and solemn ceremonies, to the worship of Jehovah, who condescended to make it the place for the special manifestation of his glory. Perhaps no structure erected either before or since is to be compared with it, for its exactly proportioned and beautiful dimensions. Its various courts and apartments were capable of holding three hundred thousand persons. It was adorned with 1,453 columns, of the finest Parian marble, twisted, sculptured, and voluted; and 2,906 pilasters, decorated with magnificent capitals. The oracle and sanctuary were lined with massive gold, adorned with embellishments in sculpture, and set with numerous gorgeous and dazzling decorations of diamonds and all kinds of precious stones. In the emphatic language of Josephus, 'The Temple shined and dazzled the eyes of such as entered it by the splendor of the gold that was on every side of them.'

"The multitude on beholding it were struck with bewildering amazement, and raised their hands in admiration and astonishment at its wondrous magnificence, as well as to protect their eyes from the effect of its exceeding brilliancy.

"Nothing ever equalled the splendor of its consecration. Israel sent forth her thousands, and the assembled people beheld, in solemn adoration, the vast sacrifice of Solomon accepted. The flame descended upon the altar and consumed the offering; the shadow and glory of the Eternal proclaimed his presence between the cherubim, and the voice of his thunders told to the faithful of the craft that the perfectness of their labor was approved.

> *   *   *   *   Bright was the hour
> When Israel's princes, in their pride and power,
> Knelt in the Temple's court: the living flame,
> The accepted sacrifice to all proclaim.
> Brightly the splendor of the Godhead shone,
> In awful glory, from His living throne;
> Then bowed was every brow—no human sight
> Could brave the splendor of that flood of light
> That veiled His presence and His awful form—
> Whose path the whirlwind is—whose breath the storm."
>
> —R. A. Monitor, p. 51-53.

The historical lecture usually ends here, or the Master may continue as in the Monitor, p. 123, (Sheville & Gould); the Candidate is requested to stand, and the Right Worshipful Master reads or repeats the usual

CHARGE.

R. W. M.: "Brother Hunt, your admittance to this degree of Masonry is a proof of the good opinion the brethren of this lodge entertain of your Masonic abilities. Let this consideration induce you to be careful of forfeiting, by misconduct and inattention to our rules, that esteem which has raised you to the rank you now possess.

"It is one of your great duties, as a Most Excellent Master, to dispense light and truth to the uninformed Mason; and I need not remind you of the impossibility of complying with this obligation, without possessing an accurate acquaintance with the lectures of each degree.

"If you are not already completely conversant in all the degrees heretofore conferred on you, remember that an indulgence, prompted by a belief that you will apply yourself with double diligence to make yourself so, has induced the brethren to accept you.

"Let it therefore be your unremitting study to acquire such a degree of knowledge and information as shall enable you to discharge with propriety the various duties incumbent on you, and to preserve unsullied the title now conferred upon you, of a Most Excellent Master."

After which the Candidate is conducted to a seat among the rest of the members.

CHAPTER VIII..

MOST EXCELLENT MASTER.

CLOSING THE LODGE.

The following is the correct manner of closing a lodge of Most Excellent Masters, the ceremonies being given in full for the special benefit of the Masonic student:

R. W. M. (one rap): "Brother Junior Deacon, what is the last as well as the first great care of assembled Masons?"

J. D.: "To see that the lodge is duly tyled, Right Worshipful."

R. W. M.: "Perform that duty. Inform the Tyler that I am about to close this lodge of Most Excellent Masters; direct him to take due notice and tyle accordingly."

The Junior Deacon gives six knocks on the anteroom door, as at opening. (See p. 134.) These are answered in the same way by the Tyler. The Junior Deacon opens the door, whispers to the Tyler, "Going to close the lodge;" closes the door, and gives three double knocks as before, which are answered by the Tyler after the same manner. The Junior Deacon then faces the East, makes the due guard and sign and reports:

J. D.: "Right Worshipful Master, the Tyler is informed and the lodge is duly tyled."

R. W. M.: "How is it tyled?"

J. D.: "By a worthy brother without, armed with the proper implement of his office."

R. W. M.: "His duty there?"

J. D.: "To keep off cowans and eavesdroppers, and admit none but such as are duly qualified and have permission from the Right Worshipful Master."

R. W. M. (one rap): "Brother Senior Warden, are you a Most Excellent Master?"

S. W.: "I am; try me."

R. W. M.: "How will you be tried?"

S. W.: "By the Keystone."

R. W. M.: "Why by the Keystone?"

S. W.: "Because at the completion and dedication of King Solomon's Temple and the placing of the Keystone, this degree was founded."

R. W. M.: "Where were you made a Most Excellent Master?"

S. W.: "In a legally constituted, and duly opened, lodge of Most Excellent Masters."

R. W. M.: "How many compose such a lodge?"

S. W.: "Two, or more."

R. W. M.: "When composed of two, who are they?"

S. W.: "The Right Worshipful Master and Senior Warden."

R. W. M.: "The Senior Warden's station?"

S. W.: "In the West."

R. W. M.: "Why in the West, and your duty there?"

S. W.: "As the sun is in the West at the close of the day so is the Senior Warden in the West to assist the Right Worshipful Master in opening and closing the lodge, to pay the Craft their wages if aught be due, harmony being the strength and support of all well governed institutions."

R. W. M.: "The Right Worshipful Master's station?"

S. W.: "In the East."

R. W. M.: "Why in the East, and his duty there?"

S. W.: "As the sun rises in the East to open and govern the day so rises the Right Worshipful Master in the East to open and govern the lodge, set the Craft to work and give them proper instruction for their labors."

R. W. M. (rising gives three raps and all the brethren stand): "Brother Marshal you will assemble the brethren around the altar and see that they are in due form for our devotions."

Marshal: "Brethren, you will assemble around the altar and be in due form for our devotions."

The brethren form a circle around the altar, leaving an open space in the East and West parts of the circle for the Right Worshipful Master and Senior Warden to fill. They then kneel on the right knee and form a chain with the right arm over the left, as in opening the lodge. (See p. 137.)

Marshal: "Right Worshipful Master, the brethren are in due form for our devotions, and await your presence."

The Right Worshipful Master and Senior Warden descend from their respective stations, and fill the spaces left for them in the circle, kneeling on the right knee, and, with the right arm over the left, complete the chain. They then repeat the Lord's prayer in concert, as in opening—balance one, two, three—one, two, three—as before. Then all rise, and the Right Worshipful Master and Senior Warden return to their stations, the brethren standing around the altar.

R. W. M.: "Brother Senior Warden it is my order that this lodge of Most Excellent Masters be now closed; this order you will communicate to the brethren present for their government."

S. W.: "Brethren, it is the order of the Right Worshipful Master that this Most Excellent Masters' lodge be now closed; of this order you will take due notice and govern yourselves accordingly. Look to the East."

All the signs are then made in concert with the Right Worshipful Master, beginning with those of Most Excellent Master, down to the Entered Apprentice degree inclusive. The R. W. M. gives two quick raps; the S. W. gives two; R. W. M. gives two more; the S. W. the same; R. W. M. gives two more and the S. W. two; each officer thus giving six raps, this degree being considered the sixth in the American series.

R. W. M. (reads from the Monitor): "The Lord is my shepherd; I shall not want. He maketh me to lie down in green pastures; he leadeth me beside the still waters. He restoreth my soul; he leadeth me in the paths of righteousness for his name's sake. Yea, though I walk through the valley of the shadow of death, I will fear no evil; for thou art with me;

thy rod and thy staff they comfort me. Thou pre-
parest a table before me in the presence of mine
enemies; thou anointest my head with oil; my cup
runneth over.   Surely goodness and mercy shall follow me all the days of my life; and I will dwell in
the house of the Lord forever."—Ps. xxiii.

R. W. M.: "I now declare this lodge of Most
Excellent Masters closed. Brother Junior Deacon,
so inform the Tyler. Brother Senior Deacon take
charge of the Great Lights."

MOST EXCELLENT MASTER LECTURE.

Following is the Lecture which the Candidate is
supposed to commit to memory, being orally instructed therein by some well-posted brother, but
which henceforth can be learned easier and more
correctly from these pages. A brother visiting a
strange lodge is also examined in this Lecture, and
hence it ought to be thoroughly memorized. Our
Candidate, Rev. James Hunt, we will suppose is being examined in open chapter, and seated in front
of the East.

Examiner: "Are you a Most Excellent Master?"

Candidate: "I am."

Ex.: "How gained you admission?"

Can.: "By six distinct knocks."

Ex.: "To what do those knocks allude?"

Can.: "To the sixth degree of Masonry, it being that upon which I was about to enter."

Ex.: "What was said to you from within?"

Can.: "Who comes here?"

Ex.: "Your answer?"

Can.: "A worthy brother, who has been regularly initiated, passed, and raised to the sublime degree of Master Mason, advanced to the honorary degree of a Mark Master, has been inducted into the Oriental Chair of King Solomon, and now seeks further promotion in Masonry by being received and acknowledged a Most Excellent Master."

Ex.: "What was then asked you?"

Can.: "If it was my own free will and accord that I made that request, if I was duly and truly prepared, worthy and well qualified, if I had made suitable proficiency in the preceding degrees; all of which being answered in the affirmative, I was asked by what further right and benefit I expected to gain admission."

Ex.: "Your answer?"

Can.: "By the benefit of the pass."

Ex.: "Had you the pass?"

Can.: "I had it not; my conductor gave it for me."

Ex.: "What was it?"

Can.: "Rab-bo-ni."

Ex.: "What was then said to you?"

Can.: "I was told to wait until the Right Worshipful Master could be informed of my request and his answer returned."

Ex.: "What was his answer?"

Can.: "Let the brother enter and be received in due form."

Ex.: "How were you received?"

Can.: "Upon the Keystone, because at the completion and dedication of the Temple, the stone which the builders rejected became the head stone of the corner."

Ex.: "How were you then disposed of?"

Can.: "I was conducted six times about the lodge to the Senior Warden in the West, and the Right Worshipful Master in the East, where the same questions were asked, and like answers returned as at the door."

Ex.: "How did the Right Worshipful Master dispose of you?"

Can.: "He ordered me to be re-conducted to the West, and placed in charge of the Senior Warden, who would teach me how to approach the East in a proper manner."

Ex.: "What is that proper manner?"

Can.: "Advancing by six upright steps; first, as an Entered Apprentice; second, as a Fellow Craft; third, as a Master Mason; fourth, as a Mark Master; fifth, as a Past Master; sixth, advancing one step with the right foot, bringing the heel of the left to the heel of the right, my feet forming a right angle, my body erect, facing the East."

Ex.: "What was then done with you?"

Can.: "I was made a Most Excellent Master."

Ex.: "How?"

Can.: "In due form."

Ex.: "What was that due form?"

Can.: "Kneeling at the altar on both knees, both hands resting on the Holy Bible, Square and Compass; in which position I took upon myself the solemn oath or obligation of a Most Excellent Master."

Ex.: "Have you that solemn oath or obligation?"

Can.: "I have."

Ex.: "Repeat it." The obligation is very seldom repeated, (p. 153.)

Ex.: "Show me a sign."

Candidate gives the due guard aud penal sign. (See figure.)

Ex.: "What is that?"

Can.: "The due guard and penal sign of a Most Excellent Master."

• *Ex.*: "To what does it allude?"

Can.: "To the penalty of my obligation."

Ex.: "Show me another sign."

Candidate gives the sign of admiration. (See figure, p. 162.)

Ex.: "What is it called?"

Can.: "The sign of admiration."

Ex.: "To what does it allude?"

Can.: "To the wonder and astonishment expressed by those of our ancient brethren who were permitted to view the interior of that magnificent edifice which King Solomon had erected, and was about to dedicate to the Supreme Being."

Ex.: "Have you any grips belonging to this degree?"

Can.: "I have."

Ex.: "Give me a grip."

Candidate gives the grip of a Most Excellent Master, pressing the thumb upon the third knuckle-joint of the examiner's right hand. (See figure.)

Ex.: "What is that?"

Can.: "The grip of a Most Excellent Master."

Ex.: "Has it a name?"

Can.: "It has."

Ex.: "Give it."

Can.: "Rab-bo-ni."

Ex.: "What is it otherwise called?"

Can.: "The covering grip, because .as this covers the grips of preceding degrees, so ought we as Most Excellent Masters, considering that man in his best estate is subject to frailty and error, endeavor to cover his faults with the broad mantle of charity and brotherly love."

END OF THE LECTURE.

The very best that can be said of the foregoing ceremonies is that they are an uncouth imitation and poorly executed even at that, while the constant repetition of Scripture passages in all these Chapter degrees is simply a solemn mockery, alike dishonoring to God, and an insult to true Christian intelligence. But looking at Masonry from any stand point we please, the question may be asked with ever increasing emphasis, what necessity is there for all those terrible oaths with their inhuman and unlawful penalties of death?

CHAPTER IX.

ROYAL ARCH INTRODUCTION.

In the Masonic ritual, and for that matter in all standard works on Masonry, the third degree is always referred to as "the sublime degree of Master Mason," but when speaking of the Royal Arch degree, it is declared to be "indescribably more august, sublime and important, than all which precede it, and is the summit and perfection of ancient Masonry." ("R A. Standard.")

Unlike the preparatory degrees, we know for a certainty by whom, when and where the Royal Arch degree was manufactured, the causes which led to its construction, the circumstances connected with its early history, and every detail relating to its adoption as a part of the great anti-christian system of Freemasonry, all these particulars being furnished us not by the enemies of Masonry, but by its ablest and most zealous advocates and teachers. In the "Traditions of Freemasonry," by A. T. C. Pierson, Past Grand Master of Minnesota, page 320, we read as follows:

"About A. D. 1740, the Chevalier Ramsay appeared in London. He was a Scotchman by birth, but had long been a resident of France; a zealous partisan of the Pretender (Charles, son of the dethroned papist King, James II.) he sought to advocate the Stuart interest by the use of Masonry. He brought with him several new degrees which he endeavored to introduce in the English lodges. Among these was one which he called Royal Arch."

Previous to the advent of Ramsay a schism of large dimensions had taken place in the ranks of Masonry in England, and quite a number of its most influential members, headed by an Irishman named Lawrence Dermott, constituted themselves into an independent Grand lodge, and established their headquarters at York.

For a number of years a most bitter feud existed between the two Grand lodges of London and York, the latter styling themselves "Ancient Masons," while they derisively applied the term "Modern" to their former associates of .London. Ramsay's pet scheme was the restoration of the Stuart family to the throne of England, and the re-establishment of popery as the religion of the country, but the London Grand lodge rejecting his new-fangled degree he was unable to carry out his project.

"The shrewd Lawrence Dermott, (Pierson informs us) who was for many years the active spirit among the seceders, saw in this new degree a means of drawing attention to the Ancient Lodges and to increase their popularity. After a time it was claimed and asserted (by the York Masons) that the 'Moderns' were ignorant of the Master's part, and that the 'Ancients' alone had that knowledge, and that there were four degrees in Ancient Masonry, whereas the (London) Grand lodge acknowledged only three."

"In the meantime a man named "Thomas Dunckerley, (as Pierson tells us) an illegitimate son of George II, was chosen Master of a regular lodge in A. D. 1770; he soon assumed a high position among the distinguished Masons of the age, and finally became Grand Master. Visiting the Ancient lodges he became acquainted with the new degree, and determined to introduce it into the regular lodges."

"Divesting the degree of many of its crudities; in fact remodeling it and revising the lecture, he presented the Holy Royal Arch Degree to the (London) Grand lodge. It was at once found that the practice of this new degree required a change in that of the Master, Mason's degree—a removal and a substitution—a transfer, says Dr. Oliver, of the Master's word." ("Traditions of Freemasonry," Pierson, pp. 320-322.)

And thus we have an authoritative statement from the leading Masonic teachers of the century as to the alleged loss of "the Master's word," and its subsequent recovery, notwithstanding the stupid fable about the death of Hiram Abiff, while we are also assured that the Royal Arch degree as we have it to-day has come down to us from clandestine lodges and through the instrumentality of the bastard son of an English king. No wonder then that it bears the name of "Royal."

This degree is a sort of dramatic representation of the destruction of the city and Temple of Jerusalem by Nebuzaradan, the Babylonian general, and the rebuilding of the second Temple under Zerubbabel at the close of the seventy years' captivity of

Judah and Benjamin. A Royal Arch Chapter therefore, when properly equipped, is supposed to represent the alleged Tabernacle erected near the ruins of the first, during the erection of the second Temple.

The officers of the Chapter, with their regalia and jewels are as follows:

1.   High Priest, whose title is "Excellent." He sits in the East, clothed in a robe of blue purple, scarlet and white, and is decorated with imitations of the ephod, breastplate and mitre, the garments and decorations of the ancient High Priest of the Jews. On the front of the mitre, upon a plate supposed to be gold, is inscribed "Holiness to the Lord." His jewel is a mitre.

2.   The King, representing Zerubbabel. His station is in the East, at the right of the High Priest. He is clothed in a scarlet robe, with a crown upon his head, and a sceptre in his hand. His jewel is a level, surmounted by a crown.

3.   The Scribe represents Haggai, the prophet. He is stationed in the East, to the left of the High Priest, clothed in a purple robe and wearing a turban of the same color. His jewel is a plumb, surmounted by a turban.

These three officers constitute the Grand Council, and are correctly represented in the figures on page 187, which are taken with slight changes from Armstrong's catalogue of Chapter goods.

4.   The Captain of the Host, stationed to the right of the Council and a little in front. He wears a white robe and turban and is armed with a sword.

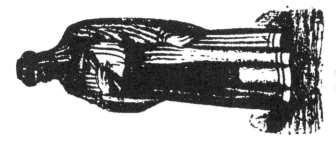

Scribe.

High Priest.

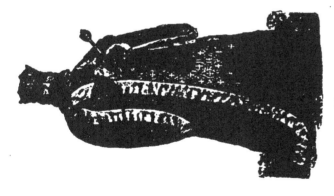

King.

His jewel is a triangular plate, on which an armed soldier is engraved.

5. The Principal Sojourner, seated on the left of the Council, and on a line with the Captain of the Host. He wears a black robe, with a rose-colored border, a turban, and carries a pilgrim's staff. His jewel is a triangular plate on which a pilgrim is engraved.

6. The Royal Arch Captain. He wears a white robe and turban, and carries a sword. His station is within the Fourth Vail, or Sanctuary. His jewel is a triangular plate, on which is engraved a sword.

7. The Master of the Third Vail wears a scarlet robe and turban. His station is within the Third Vail, the color of which is scarlet. His jewel is a sword.

8. The Master of the Second Vail wears a purple robe and turban. He sits within the Second Vail, the color of which is purple. His jewel is a sword.

9. The Master of the First Vail wears a blue robe and turban. His station is within the First Vail, and his jewel is a sword.

The Treasurer, Secretary and Sentinel occupy their respective places as in previous degrees. and wear their appropriate jewels. For the stations of all these officers see diagram, page 226.

The apron of a Royal Arch Mason is white lambskin, lined and bound with scarlet: on the flap of which is a triple tau within a triangle, and all within a circle, as in the annexed figure.

The collar and sash are scarlet, edged with gold·
or yellow material, the sash being worn from the left
shoulder to the right hip.' In conferring the preparatory degrees, from the Mark Master up.. the
apron and collar of the Chapter are worn by the
companions.

In the Royal Arch Degree they ought also to
wear white robes, simply made, with a draw-string

at the throat and waist, the officers, of course, wearing their own appropriate robes. The costumes of the officers are not the same in all Chapters, being differently made by different firms, but those in most common use are correctly represented in the diagrams presented in this degree.

OPENING CEREMONIES.

The officers and companions being clothed, the High Priest takes his station, and calls to order by one rap with his gavel. All are seated as in diagram, page 199.

Excellent High Priest: "Companion Captain of the Host, are all present Royal Arch Masons?"

If satisfied that they are, he replies:

Captain of the Host: "All present are Royal Arch Masons."

But if not so satisfied, he answers:

C. of H.: "I will ascertain and report."

C. of H.: "Companion Royal Arch Captain, you will ascertain if all present are Royal Arch Masons."

The Royal Arch Captain beginning near his own station proceeds around the room, each companion as he approaches rising to his feet and whispering the pass—"I am that I am"—in his ear. Having thus collected the pass from all but the principal officers, he repairs to his proper place and reports:

Royal Arch Captain: "Companion Captain of the Host, all present are Royal Arch Masons," and takes his seat.

C. of H. (reports): "Excellent High Priest, all present are Royal Arch Masons."

This ceremony of "purging" being gone through with, the due guard and sign must always be made as a salute in addressing a superior officer, or when entering or retiring from the Chapter.

E. H. P.: "Are there a constitutional number present to open a Chapter of Royal Masons?"

C. of H.: "There are three times three."

E. H. P.: "You will take the steps preliminary to opening a Chapter."

C. of H.: "Companion Royal Arch Captain, when a Chapter of Royal Arch Masons is about to be opened, what is your duty?"

R. A. C.: "To see that the Sentinel is at his post, and the Tabernacle securely guarded."

C. of H.: "Perform that duty, and inform the Sentinel that we are about to open a Chapter of Royal Arch Masons, and direct him to guard accordingly."

The Royal Arch Captain goes to the ante-room door, opens it and whispers to the Sentinel: "Going to open the Chapter." He then closes the door and gives seven knocks: two—two—three, which are answered in the same manner by the Sentinel on the outside. The R. A. C. returns to his station and reports:

R. A. C.: "Companion Captain of the Host, the Sentinel is at his post and the Tabernacle is securely guarded."

C. of H.: "Excellent High Priest, your orders have been duly executed."

E. H. P.: "Companion Captain of the Host, are you a Royal Arch Mason?"

C. of H.: "I am that I am." *

E. H. P.: "How shall I know you to be a Royal Arch Mason?"

C. of H.: "By three times three under a living arch and over a triangle."

E. H. P.: "Why in that manner?"

C. of H.: "Because in that manner only can the principal secrets of this degree be communicated."

E. H. P.: "Where were you made a Royal Arch Mason?"

C. of H.: "In a legally constituted Chapter of Royal Arch Masons, assembled in a place representing the Tabernacle erected by our ancient brethren near the ruins of King Solomon's Temple."

E. H. P.: "How many compose a Chapter of Royal Arch Masons?"

C. of H.: "Nine, or more."

E. H. P.: "When composed of only nine, who are they?"

C. of H.: "The Excellent High Priest, Companions King and Scribe, Captain of the Host, Principal Sojourner, Royal Arch Captain, and the three Masters of the Vails."

E. H. P.: "Whom do the first three represent."

C. of H.: "Joshua, Zerubbabel, and Haggai, who composed the first Grand Council assembled at Jerusalem after the destruction of the first Temple, and held their meetings in the Tabernacle."

E. H. P.: "Whom do the latter three represent?"

---

* This is a mere bandying of Jehovah's name, yet both bishop and preacher not only sanction but actually swear to maintain and support this solemn mockery.

C. of H.: "Those three of our ancient brethren who were instrumental in bringing to light the principal secrets of this degree, after they had lain buried in darkness from the death of our Grand Master Hiram Abiff until the building of the second Temple, a period of 470 years, and for their valuable services were appointed Masters of the Vails."

E. H. P.: "How many vails were there?"

C. of H.: "Four."

E. H. P.: "To what do they allude?"

C. of H.: "To the principal tribes of Israel which bore their banners in the wilderness—Judah, Ephraim, Reuben and Dan, whose emblems were the Lion, the Ox, the Man and the Eagle."

E. H. P.: "Where were the vails placed?"

C. of H.: "At the outer courts of the Tabernacle."

E. H. P.: "For what purpose?"

C. of H.: "To serve as a covering for the Tabernacle and stations for the guards."

E. H. P.: "Why were guards placed?"

C. of H.: "To see that none passed but such as were duly qualified and had permission, none being admitted to the presence of the Excellent High Priest, King and Scribe sitting in council, except the true descendants of the twelve tribes of Israel."

E. H. P.: "What do these banners emblematically teach?"

C. of H.: "That when engaged in the pursuit of truth—the great object of Masonic study—we should have the courage of the lion, the patience of the ox, the intelligence of the man, and the swiftness of the eagle."

E. H. P.: "The station of the Master of the First Vail?"

C. of H.: "Within the First Vail."

E. H. P. gives one rap, and the Master of the First Vail rises.

E. H. P.: "Your duty, companion?"

Master First Vail: "To guard the First Vail and admit none but such as are duly qualified and have the pass."

E. H. P.: "What is the pass?"

M. 1st V.: "I am that I am."

E. H. P.: "What is the color of your Vail?"

M. 1st V.: "Blue, emblematical of friendship, and peculiarly characteristic of a Master Mason."

E. H. P.: "The station of the Master of the Second Vail?"

M. 1st V.: "Within the Second Vail."

E. H. P. (one rap, calling up Master of the Second Vail): "Your duty, companion?"

Master Second Vail: "To guard the Second Vail and admit none but such as are duly qualified, and have the words, sign and words of explanation of the Master of the First Vail."

E. H. P.: "What are the words?"

M. 2nd V.: "Shem, Ham and Japhet."

E. H. P.: "What is the color of your Vail?"

M. 2nd V.: "Purple, which being formed by a due admixture of blue and scarlet, is therefore placed between the First and Third Vails, which are of these colors, to remind us of that intimate connection that exists between Symbolic Masonry and the Royal Arch degree."

E. H. P.: "The station of the Master of the Third Vail?"

M. 2nd V.: "Within the Third Vail."

E. H. P. (one rap, Master of the Third Vail rises): "Your duty, companion?"

Master Third Vail: "To guard the Third Vail and admit none but such as are duly qualified and have the words, sign, and words of explanation of the Master of the Second Vail."

E. H. P.: "What are the words?"

M. 3rd V.: "Moses, Aholiab and Bazaleel."

E. H. P.: "What is the color of your Vail?"

M. 3rd V.: "Scarlet, emblematic of that fervency and zeal which should actuate all Royal Arch Masons, and is peculiarly characteristic of this degree."

E. H. P.: "The station of the Royal Arch Captain?"

M. 3rd V.: "Within the Fourth Vail, or entrance to the Sanctuary."

E. H. P. (one rap, and that officer rises): "Your duty, companion?"

R. A. C.: "To guard the Fourth Vail and admit none but such as are duly qualified and have the words, sign, and words of explanation of the Master of the Third Vail and the signet of truth."

E. H. P.: "What are the words?"

M. 4th V.: "Zerubbabel, Joshua and Haggai."

E. H. P.: "What is the color of your Vail?"

M. 4th V.:   "White, emblematical of that purity of life and rectitude of conduct by which alone we can expect to gain admission into the Holy of Holies above.".

E. H. P.:   "The station of the Principal Sojourner."

M. 4th V.:   "At the left of the Council."

E. H. P. (one rap, and Principal Sojourner rises):   "Your duty, companion?"

Principal Sojourner: "To bring the blind by a way that they knew not, to lead them in paths that they have not known, to make darkness light before them and crooked things straight; these things do unto them, and not forsake them."

E. H. P.:   "The station of Captain of the Host?"

P. S.:   "At the right of the Council."

E. H. P. (one rap, and the Captain of the Host stands):   "Your duty, companion?"

C. of H.:   "To observe the orders of the Excellent High Priest and see them duly executed, to take charge of the Chapter during the hours of labor, and superintend the introduction of strangers among the workmen."

E. H. P.:   "The station of the Scribe?"

C. of H.:   "On the left in Council."

E. H. P.:   "His duty?"

C. of H.:   "To assist the Excellent High Priest in the discharge of his duties, and in his absence, and that of the King, to preside over the Chapter."

E. H. P.:   "The station of the King?"

C. of H.:   "On the right in Council."

E. H. P.:   "His duty?"

C. of H.: "To assist the Excellent High Priest in the discharge of his duties, and in his absence to preside over the Chapter."

E. H. P.: "The station of the Excellent High Priest?"

Due Guard.   Grand Hailing Sign.

C. of H.: "In the East, in the center of the Council."

E. H. P.: "His duty?"

C. of H.: "To preside over and govern the Chapter with fidelity, read and expound the law, officiate in the Temple, and offer up the incense of a pure and contrite heart to the Great I Am."

E. H. P.: "Companion Captain of the Host, you will bring the companions to order as Royal Arch

Masons, and assemble them around the altar for our devotions."

C. of H.: (gives three raps, and all rise): "Companions, you will come to order as Royal Arch Masons."

All the companions, standing, make the due guard of a Royal Arch Mason, as seen in the figure p. 197—edge of the right palm to the forehead, as if guarding the eyes from the light of the sun.

C. of H.: "You will assemble around the altar for our devotions."

All except the High Priest, King and Scribe assemble in a circle around the altar, leaving a space open for the Council and Captain of the Host.

C. of H.: "Excellent High Priest, your orders have been executed."

The Council rise to their feet, and the High Priest reads the following

CHARGE.

"Now we command you, brethren, that ye withdraw yourselves from every brother that walketh disorderly, and not after the tradition which he received of us. For yourselves know how ye ought to follow us; for we behaved ourselves not disorderly among you. Neither did we eat any man's bread for nought, but wrought with labor and travail day and night that we might not be chargeable to any of you. Not because we have not power, but to make ourselves an ensample unto you to follow us. For even when we were with you, this we commanded

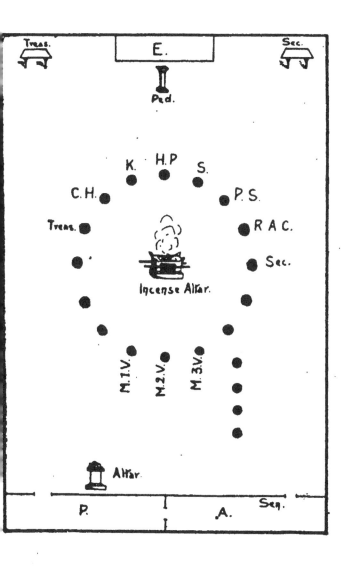

you, that if any would not work, neither should he eat; for we hear there are some who walk among you disorderly, working not at all, but are busybodies. Now them that are such, we command and exhort, that with quietness they work, and eat their own bread. But ye, brethren, be not weary in well doing. And if any man obey not our word, note that man, and have no company with him, that he may be ashamed. Yet count him not as an enemy, but admonish him as a brother. Now the Lord of peace himself give you peace always." (2 Thess. iii: 6-16.) Mackey's Ritualist, page 348.

The High Priest, King and Scribe descend, and with the Captain of the Host fill up the vacant spaces in the circle. All then kneel on the right knee and form a chain with their right arms over the left, as in the Most Excellent Master's degree, the relative positions of the officers and companions being correctly represented in the diagram on page 199. The High Priest, Chaplain, or preacher (if one be present) now repeats or reads the following alleged prayer from the Monitor:

"Direct us, O Lord, in these and all our doings, with thy most gracious favor, and further us with thy continual help, that in all our works begun, continued and ended in thee, we may glorify Thy holy name, and finally, by thy mercy, obtain everlasting life. Amen."

Companions (respond): "So mote it be." (R. A. Standard, p. 65.)

Or they may repeat the Lord's Prayer in concert with the High Priest, as in a preceding degree.

After prayer, and still kneeling around the altar and maintaining the chain, they balance three times three, that is they raise their arms up, down, up, etc., nine times, at the end of which the Excellent High Priest says: "Arise," and all rise to their feet, breaking the chain.

E H P.: "Companion Captain of the Host, you will form the companions in groups of three commencing at the right."

C. of H.. "Companions, you will form yourselves into groups of three."

The Captain of the Host sees that the groups are formed as follows: Each companion with his

right hand grasps the right wrist of the companion
on his left, and with his left hand he grasps the left
wrist of the companion on his right, the left arm being under the right.   Each companion of a group
then places the toe of his right foot to the heel of
the right foot of the companion on his right, their
feet thus forming a triangle.   The High Priest, King
and Scribe form one group, as do the Masters of the
Vails.

C. of H.: "Excellent High Priest, the groups are
formed."   See figure p. 201.)

E. H. P.:—.

"As we three did agree
In peace, love and unity,
The sacred word to search :

"As we three did agree
In peace, love and unity,
The sacred word to keep :

"So we three do now agree
In peace, love and unity,
To raise a Royal Arch."

The groups then give three times three knocks,
with the right hands over the left and clasped as
explained above.   The right arms are now raised,
the wrists being still grasped, and the Grand Omnific
Royal Arch Word is communicated from right to
left, in syllables, by each group, each companion, in
succession, beginning the first syllable. so that the
word is given three times, thus:

1st Comp.	*2nd Comp.*	*3rd Comp.*
Jah	Bel	On
	Jah	Bel
On		Jah
Bel	On.	

hardships and dangers in each, but I think, all things considered, the river route is more desirable. Let us pursue it."

They rise up and begin the first circuit, moving quite slowly towards the "rough road" laid down along the south side of the room, and the

P. S. (continuing): "The recent inundations of the valleys seem to have brought trees or limbs and other obstructions down from the mountains and deposited them directly across our pathway. There seems to be no choice but to go over some and under those which have lodged higher. It is a dangerous pass; but stop—have we forgotten that our trust is in the great I Am? Let us first invoke His aid and protection."

They are now near the beginning of the "rough road" where they kneel and the Principal Sojourner recites the following:

P. S.: "Lord, I cry unto thee: make haste unto me; give ear unto my voice. Let my prayer be set forth before thee, as incense: and the lifting up of my hands as the evening sacrifice. Set a watch, O Lord, before my mouth; keep the door of my lips. Incline not my heart to any evil thing, to practice wicked works with men that work iniquity. Let the righteous smite me; it shall be a kindness: and let him reprove me; it shall be an excellent oil. Mine eyes are unto thee, O God the Lord; in thee is my trust; leave not my soul destitute. Keep me from the snare which they have laid for me, and the gins of the workers of iniquity Let the wicked fall into their own nets, whilst that I withal escape." (Ps. cxli.) "Arise, let us be going."

Rising up they enter upon the "rough and rugged road," over which they travel with great difficulty, as represented on page 235, being often jostled hither and thither by some of the companions who hang on to the connecting rope and increase the hardship of the journey, the Principal Sojourner reciting as they proceed:

P. S.: ["How it lightens the burdens, how the dangers fade, as it were, when we in sincerity and truth call upon God. Yet He has given us sinews for use, a mind and will which are to be exercised under His guidance.] Be careful, stoop beneath that limb, now over this fallen tree. Ah! we are safely passed. It was like a lesson of trust, effort and humility. I will bring the blind by a way that they knew not; I will lead them in paths that they have not known, I will make darkness light before them and crooked things straight. These things will I do unto them and not forsake them."

P. S.: "How beautiful is the scenery along the banks of this famous Euphrates! It rises in the mountains of Armenia and flows westerly; but for an intervening range of hills it would flow into the Mediterranean. Then it changes its course and empties into the Erythrian Sea. The river Tigris finds its course very near the headwaters of the Euphrates, and at once takes a south-easterly course, supplies Nineveh with water, approaches to within fifteen miles of the Euphrates and coquettishly flows away to unite with it before it reaches the sea. Be-

PASSING OVER THE ROUGH ROAD.

tween these rivers is the country—almost an im-
mense island—called Mesopotamia, the birthplace
of our father Abraham."

(Second Circuit.) "We have left the river and
are traversing the Syrian desert, which borders the
one we avoided. No water anywhere! Oh, see the
beautiful foliage in the distance! It is Tadmor in
the wilderness. It is a most delightful spot. Does
it not look like Babylon, with its towers and Corin-
thian columns and domes? [All this country was
conquered by King David, and Tadmor was built by
King Solomon, who inherited the kingdom from the
Euphrates to the Mediterranean, and from the
mountains of Lebanon to the Red Sea. Its beautiful palm trees, pure water, and the munificence of
King Solomon have made this oasis in the desert to
blossom as the rose. Caravans stop here while on
their long journeys. But Tadmor will become a
heap of ruins, desolate and forsaken by Him whom
its people forsake. The Lord is true to those who
serve him, but those who forsake him shall miser-
ably perish.] What is that I see? Two caravans
engaged in combat! Let us go around and avoid
them."

Clashing of swords is heard, and they make a
slight detour, or passing into the ante-room and
through to the Preparation room door by which they
again enter the Chapter.

P. S.: "That was a narrow escape, and it has
brought us into an abandoned path. The old
bridge over this ravine appears to be unsafe. Stop,
let me examine it."

They halt and he goes forward a few steps as if to make an examination, and returning says:

P. S.: "It is nearly gone, but by care we may cross it. Let us again call upon the God of our fathers."

Candidates kneel again, this time a little inside the Preparation room door, and the Principal Sojourner recites the second prayer. (R. A. Standard, p. 74.)

P. S.: "I cried unto the Lord with my voice; with my voice unto the Lord did I make my supplication. I poured out my complaint before him; I showed before him my trouble. When my spirit was overwhelmed within me, then thou knewest my path. In the way wherein I walked have they privily laid a snare for me. I looked on my right hand, and beheld, but there was no man that would know me; refuge failed me: no man cared for my soul. I cried unto thee, O Lord; I said, Thou art my refuge and my portion in the land of the living. Attend unto my cry, for I am brought very low; deliver me from my persecutors; for they are stronger than I. Bring my soul out of prison, that I may praise thy name." —Ps. cxlii. "Arise, brethren, and let us make the attempt."

They now move on over the "old bridge," usually set up along the north side of the room. It is about as long as "the rough road," and threatens to fall with every step they take. (See p. 238.)

P. S.: "Be careful, the bridge totters, but we will soon be safely over. It seems ready to fall. Be

CROSSING THE OLD BRIDGE.

careful, be careful — (the old bridge falls.) Oh!
Thank God we are safely over."

It ought to be mentioned that the "rough road"
and the "old bridge" are not uniformly laid down as
above described, each Chapter regulating such matters to suit its own convenience, but neither of
these difficult places must ever be dispensed with.

P. S.: "More desert and dreary waste, but we
will soon reach Damascus, a famous resting-place
for travelers. Yonder is Riblah, where Nebuchadnezzar tarried during the siege of Jerusalem, whea
the city was taken about midnight, and the enemy
entered into the Temple. King Zedekiah, with his
wives, children, chief captains and friends, fled
through the fortified ditch, but were overtaken near
Jericho. Many of his friends forsook him, to make
their own escape, but he was taken before Nebuchadnezzar, who rebuked him as a covenant breaker,
caused his sons and his friends to be slain, while
Zedekiah looked on; after which he put out the
king's eyes, and carried him in chains to Babylon.

"At Riblah, too, Sereiah the High Priest and
other priests and rulers of the Temple were be-
headed by order of Nebuchadnezzar; there the holy
vessels were taken, and from thence they, with
Jehosedech the High Priest, son of Sereiah, and
King Zedekiah were carried to Babylon.

"I see vegetation and signs of human habitation.
We are surrounded by vineyards and cool fountains.
Just before us is the beautiful city of Damascus,
founded by Uz, the son of Aram. It is noted for its
wealth and magnificence, and was a city of some

consequence in the time of our father Abraham. See the distant mountains, with their bright, crowning summits. Here is the Abana and before us the Pharpar. Damascus was conquered by King David after he had subdued Jerusalem, which he made his capital. [Here is a band of minstrels, let me induce them to sing some of their beautiful songs."]

They sit down as before and there is music.

P. S.: "That was delightfully sweet. Let us be going. Now we journey to the South. To the right we may see the Lebanon Mountains, and the forests where our fathers felled and prepared the timbers for the first Temple. Now we approach the Jordan. Here is Succoth, and yonder is Zarathan. Between them, in these clay grounds, is where the larger vessels of the Temple and the famous Jachin and Boaz were cast. Our father Jacob made booths at Succoth, from which it derives its name. [There our brethren camped after they had left Rameses. Below Zarathan is where the water stood in heaps as Joshua passed over.] Here is the ford leading to Jericho, that fell before Joshua's conquering hosts. Before attempting its passage, let us once more invoke the blessings of Deity. [We are weary, and although borne up by the nearness of our journey's end, we feel almost overwhelmed and desolate lest the enemy and robbers of this infested route should at the last cut off all hope of final success."]

Candidates are halted by pretended robbers, and a sham fight occurs, from which they escape. They then kneel for the third time, while the Principal Sojourner reads or recites:

P. S.: "Hear my prayer, O Lord; give ear to my supplication. In thy faithfulness answer me, and in thy righteousness.   And not enter into judgment with thy servant; for in thy sight shall no man living be justified.   For the enemy hath persecuted my soul; he hath made me to dwell in darkness.   Therefore is my spirit overwhelmed within me; my heart within me is desolate.   Hear me speedily, O Lord; my spirit faileth; hide not thy face from me, lest I be like them that go down into the pit.   Cause me to hear thy loving kindness in the morning; for in thee do I trust: cause me to know the way wherein I should walk; for I lift up my soul unto thee.   Teach me to do thy will, for thou art my God: bring my soul out of trouble, and of thy mercy cut off mine enemies, for I am thy servant."—Ps. cxliii.   (R. A. Standard.)

The vails are now drawn as in diagram page 226, and a shallow trough-like receptacle containing enough water to cover the bottom is laid down where most convenient; the Candidates' slippers are removed, and the Principal Sojourner continues:

P. S.: "Now we enter the Jordan; its waters are not cold."   (The Candidates are led through the shallow water, their feet dried with a towel, and slippers replaced.)

"We are safely over, and near to our beloved Jerusalem.   Alas! the ruins that we behold.   Our beautiful Temple leveled with the dust, the Ark and the holy vessels gone; but cheer up; we shall give our aid to restore the waste places and engage in the noble and glorious purpose of rebuilding our city and Temple. Rough, rugged and dangerous has been our road, long and tedious our journey, but sustained and favored by the Great I Am, we have at last arrived at our journey's end. I see a tabernacle just before us. Look."

They are going towards the First Vail, are halted and their hoodwinks removed.

P. S.: "Let us at once endeavor to gain admission." Approaching the 1st Vail the Principal Sojourner gives six knocks on the floor with his staff—** ** **

M. 1st V.: "Who dares approach this 1st Vail of our sacred Tabernacle?" (Gives three knocks.) "Assemble companions, the enemy is at hand."

The other Masters of the Vails come quickly forward, and the Master of the 1st Vail steps outside the 1st Vail.

M. 1st V.: "Who comes here?"

P. S.: "Three weary sojourners who have come up to help, aid and assist in rebuilding the House of the Lord without the hope of fee or reward."

M. 1st V.: "Whence came you?"

P. S.: "From Babylon."

M. 1st V.: "By an order of the council now in session, issued in consequence of disturbance having arisen from the introduction of strangers among the workmen, none are permitted to engage in this noble and glorious work except the true descendants of the twelve tribes of Israel; you will therefore be careful in tracing your genealogy. Who are you?"

P. S.: "We are of your own brethren and kin, children of the captivity, true descendants of those

noble Giblimites sent hither at the building of the
first Temple. We have been regularly initiated,
passed and raised to the sublime degree of Master
Mason, advanced to the honorary degree ot Mark
Master, have been inducted into the Oriental chair,
and received and acknowledged Most Excellent
Masters. We were also present at the destruction
of the first Temple by Nebuzaradan, by whom we
were carried away captives to the King of Babylon,
where we have remained subject to him and his successors until the reign of Cyrus, King of Persia, by
whose recent proclamation we have been liberated,
and have now come up to help. aid and assist in this
noble and glorious work."

M. 1st V.: "By what further right or benefit do
you expect to gain admission?"

P. S.: "By the will of him who has sent us."

M. 1st V.: "Who hath sent you?"

P. S.: "The God of your fathers."

M. 1st V.: "What is his name?"

P. S.: "I Am that I Am. I Am hath sent us
unto you."

M. 1st V.: "Enter this 1st Vail of our most
sacred Tabernacle."

The Masters of the other Vails return to their
stations; the cable-tows are now removed—the hoodwinks were already removed--and they are conducted within the 1st Vail by the P. S.

M. 1st V.: "Three Most Excellent Masters you
must have been thus far, to have come to promote
this noble and glorious work; but further you cannot
go without my words, sign and words of explanation.

My words are Shem, Ham, and Japhet. My sign is this:" casts a serpentine rod on the floor and picks it up by the smaller end (see cut) "and is in imitation

of the sign given by the Lord unto Moses when He commanded him to cast his rod upon the ground." My words of explanation are explanatory of that sign, and are to be found in the writings of Moses as follows:

"And Moses answered and said, But behold, they will not believe me, nor hearken unto my voice: for they will say, The Lord hath not appeared unto thee. And the Lord said unto him, What is that in thine hand? And he said, A rod. And he said. Cast it on the ground: and he cast it on the ground, and it became a serpent, and Moses fled from before it. And the Lord said unto Moses, Put forth thine hand, and take it by the tail. And he put forth his hand and caught it, and it became a rod in his hand. That they may believe that the Lord God of their fathers, the God of Abraham, the God of Isaac, and the God of Jacob, hath appeared unto thee." (Royal Arch Standard, p. 78.)

"Your alarm in the future will be seven distinct knocks, and you have my permission to pass."

Advancing towards the entrance to the Second Vail, the Principal Sojourner gives seven knocks— three—three and one: *** *** *

M. 2nd V.: "Who dares approach this 2nd Vail of our sacred Tabernacle?" (Stepping outside, he continues:) "Who comes here?"

P. S.: "Three weary sojourners, who have come up to help, aid and assist in rebuilding the House of the Lord, without the hope of fee or reward."

M. 2nd V.: "By what particular right or benefit do you expect to gain admission?" ·

P. S.: "By the words, sign, and words of explanation of the Master of the First Vail."

M. 2nd V.: "What are his words?"

P. S.: Shem, Ham and Japhet."

M. 2nd V.: "What is his sign?"

P. S.: "It is this"—dropping his staff upon the floor, and picking it up by the small end—"and is in imitation of the sign given by the Lord unto Moses, when he commanded him to cast his rod upon the ground "

M. 2nd. V.: "What are his words of explanation?"

P. S.: "Explanatory of the sign, and are to be found in the writings of Moses, as follows: 'And Moses answered and said, But behold they will not believe me, nor hearken unto my voice, for they will say, The Lord hath not appeared unto thee; And the Lord said unto him, What is that in thine hand? And he said, A rod; and He said, Cast it on the ground; and he cast it on the ground and it became a serpent, and Moses fled before it. And the Lord said unto Moses; Put forth thine hand and take it by the tail; and he put forth his hand and caught it, and it became a rod in his hand.'" (See figure.)

·In giving this sign, the person casting the rod or staff on the floor should shrink back with a slight motion of fear.

M. *2nd V.*:　"Enter this 2nd Vail of our sacred Tabernacle." The team and Principal Sojourner enter the 2nd Vail.

M. *2nd V.*:　"Three Most Excellent Masters you must have been, thus far to have come to promote this noble and glorious work; but further you cannot go without my words, sign, and words of explanation. My words are Moses, Aholiab and Bazaleel. My sign is this" thrusts his right hand into his vest bosom—"and is in imitation of the sign given by the Lord unto Moses when he commanded him

to put his hand into his bosom. My words of explanation are explanatory of that sign, and are to be found in the writings of Moses as follows: 'And the Lord said furthermore unto him, Put now thine hand into thy bosom. And he put his hand into his bosom: and when he took it out behold his hand was leprous as snow. And He said, Put thine hand into thy bosom again. And he put his hand into his bosom again; and plucked it out of his bosom, and behold it was turned again as his other flesh. And it shall come to pass, if they will not believe thee, neither hearken to the voice of the first sign, that they will believe the voice of the latter sign.' " (Ex. IV.: 6-8, R. A. Standard.)

M. 2nd V.: (continuing)—"You have my permission to pass."

Advancing to the 3rd Vail with the candidates, the Principal Sojourner again makes the usual alarm —seven knocks—three—three—one (*** *** *.)

M. 3rd V.: "Who dare approach this 3rd Vail of our sacred Tabernacle?" Steps outside. "Who comes here?"

P. S.: "Three weary sojourners, who have come up to help, aid and assist in rebuilding the House of the Lord, without the hope of fee or reward."

M. 3d V.: "By what particular right or benefit do you expect to gain admission?"

P. S.: "By the words, sign, and words of explanation of the Master of the 2nd Vail."

M. 3d V.: "What are his words?"

P. S.: "Moses, Aholiab and Bazaleel."

M. 3d V.: "What is his sign?"

P. S.: "It is this"—thrusting his hand twice into his bosom—"and is in imitation of the sign given by the Lord unto Moses when He commanded him to put his hand into his bosom."

M. 3d V.: "What are His words of explanation?"

P. S.: "Explanatory of the sign, and are to be found in the writings of Moses as follows: 'And the Lord said furthermore unto him, Put now thine hand into thy bosom. And he put his hand into his bosom; and when he took it out behold his hand was leprous as snow. And He said, Put thine hand into thy bosom again. And he put his hand into his bosom again, and plucked it out of his bosom, and behold it was turned again as his other flesh.'"

M. 3d V.: "Enter this 3d Vail of our sacred Tabernacle." (Candidates and Principal Sojourner enter inside the 3d Vail.) "Three Most Excellent Masters you must have been, thus far to have come to promote this noble and glorious work: but further you cannot go without my words, sign, and words of explanation, and the signet of truth.   My words are Joshua, Zerubbabel and Haggai.   My sign is this"— pours water from a glass or cup on the floor *—"and is in imitation of the sign given by the Lord unto

Moses when He commanded him to take of the water of the river and pour it upon the dry land. My words of explanation are explanatory of the sign, and are to be found in the writings of Moses, as follows: 'And it shall come to pass if they will not believe also these two signs, neither hearken unto thy voice, that thou shalt take of the water of the river and pour it upon the dry land, and the water which thou takest out of the river shall become blood upon the dry land.' This is the signet of truth or Zerubbabel's signet"--handing the Principal Sojour-

ner a finger-ring surmounted by a circle enclosing a triangle. (See figure.) "You have my permission to pass."

They pass on to the entrance of the 4th Vail, where the Princi-

* This sign is usually made outside the Chapter, by placing the fingers and thumb of the right hand as if holding a glass, and turning the hand inward, as if in the act of spilling something.

pal Sojourner gives the usual alarm—seven knocks. (*** *** *)

R. A. C.: "Who dare approach this 4th Vail of our sacred Tabernacle, where incense burns on our holy altar both day and night?" (Stepping outside, he continues:) "Who comes here?"

P. S.: "Three weary sojourners, who have come up to help, aid and assist in rebuilding the House of the Lord, without the hope of fee or reward."

R. A. C.: "Whence came you?"

P. S.: "From Babylon."

R. A. C.: "By an order of the Council now in session, issued in consequence of disturbance having arisen from the introduction of strangers among the workmen, none are permitted to engage in this noble and glorious work except the true descendants of the twelve tribes of Israel. You will, therefore, be careful in tracing your genealogy. Who are you?"

P. S.: "We are of your own brethren and kin, children of the captivity. We have been received and acknowledged Most Excellent Masters and as such have made ourselves known to the guards, and now want permission to appear before the Council?"

R. A. C.: "By what particular right or benefit do you expect to gain admission?"

P. S.: "By the words, sign, and words of explanation of the Master of the 3rd Vail and this signet."

R. A. C.: "What are his words?"

P. S.: "Joshua, Zerubbabel and Haggai."

R. A. C.: "What is his sign?"

P. S.: "It is this," (gives the sign: See cut, p. 249,) "and is in imitation of the sign given by the Lord unto Moses when He commanded him to take of the water of the river and pour it upon the dry land.?"

R. A. C.: "What are his words of explanation?"

P. S.: "They are explanatory of the sign, and are to be found in the writings of Moses, as follows: 'And it shall come to pass if they will not believe also these two signs, neither hearken unto thy voice, that thou shalt take of the water of the river and pour it upon the dry land, and the water which thou takest out of the river shall become blood upon the dry land.'"

R. A. C.: "Present the signet."—The P. S. hands it over.—"You will wait until the Captain of the Host is informed of your request, and his answer returned."

The Royal Arch Captain enters within the 4th Vail, and saluting the Captain of the Host makes his report.

The pot of incense on the altar is now lighted, so that when the Candidates enter the 4th Vail they may see it burning upon the altar in front of the Council.

R. A. C.: "Companion Captain of the Host, there are without three weary sojourners who have come up from Babylon to help, aid and assist in the noble and glorious work of rebuilding the House of the Lord, without the hope of fee or reward. They claim to be Most Excellent Masters, and as such

have made themselves known to the guards, and now
want permission to appear before the Council."

C. of H.: "By what particular right or benefit
do they expect to gain admission?"

R. A. C.: "By the words, sign and words of explanation of the Master of the 3d Vail, and the signet
of truth."

C. of H.: "Present the signet." (Royal Arch
Captain advances, hands the signet to Captain of
the Host, and retires to his station.)

C. of H. (saluting the E. H. P.): "Excellent
High Priest, there are without three weary sojourners who have come up from Babylon to help, aid
and assist in the noble and glorious work of rebuilding the House of the Lord, without the hope of fee
or reward. They claim to be Most Excellent Masters, and as such have made themselves known to
the guards, and now want permission to appear before the Council."

E. H. P.: "By what particular right or benefit
do they expect to gain admission?"

C. of H.: "By the words, sign, and words of explanation of the Master of the 3d Vail, and the
signet of truth."

E. H. P.: "Present the signet."

The Captain of the Host hands the signet to the
High Priest, who passes it first to the King and then
to the Scribe; after which all three pretend to consult for a few seconds, and the High Priest reads
from the Monitor, or repeats, Hag. II. 1, 4, 23.

E. H. P. "In the seventh month, in the one and
twentieth day of the month, came the word of the

Lord by the prophet Haggai, saying: 'Speak now to Zerubbabel, the son of Shealtiel, governor of Judah, and to Joshua, the son of Josedec the High Priest, and to the residue of the people, saying, Who is left among you that saw this house in her first glory? and how do ye see it now? Is it not in your eyes in comparison of it as nothing? Yet now be strong, O Zerubbabel, saith the Lord, and be strong, O Joshua, son of Josedec the High Priest, and be strong all ye people of the land, saith the Lord, and work, for I am with you, saith the Lord of Hosts. In that day, saith the Lord of Hosts, will I take thee, O Zerubbabel, my servant, the son of Shealtiel, saith the Lord, and make thee as a signet, for I have chosen thee, saith the Lord of Hosts.'" (Royal Arch Standard, p. 81.)

E. H. P. (to Captain of the Host): "The Council recognizes this" (exhibiting the signet ring) "as Zerubbabel's signet, the signet of truth; it is my order that the three brethren be admitted."

C. of H. (addressing the Royal Arch Captain): "You will admit them."

R. A. C. (to Candidates): "It is the order of the Captain of the Host that you be admitted."

The C. of H. stands at the north side of the entrance to the 4th Vail, and the R. A. C. at the south side. The Candidates are conducted into the Sanctuary by the C. of H., who places them standing in front of the alleged altar of incense, facing the Council in the East, while he himself takes up a position at their left and P. S. at their right.

E. H. P. (To Candidates): "Having represented yourselves as Most Excellent Masters, what evidence can you present to satisfy the Council that you are such?"

P. S.: "We can give the words and signs of the several degrees through which we have passed."

E. H. P.: "You may omit the words and give the signs."

The Candidates, following the lead of the P. S., now give all the signs from Entered Apprentice to Most Excellent Master, inclusive; after which the Council again pretend to consult, and the High Priest continues:

E. H. P. "Brethren the Council are satisfied that you are Most Excellent Masters, and accept with pleasure your proffered assistance. What part of the work are you willing to undertake?"

P. S.: "Any part, Excellent High Priest, even the most servile, to promote so noble and glorious a work."

The Council consult again; after which the

E. H. P. (addresses them): "From the specimen of skill you have exhibited, the Council are satisfied of your ability to perform any part, even the most difficult, but as it is necessary to remove some more of the rubbish from the easternmost part of the ruins, in order to lay the foundation of the Second Temple, you will commence your labors there; and you will be careful to observe and preserve everything of importance, as the Council have no doubt that many valuable models of excellence lie buried there, which if brought to light would prove of essential service to the Craft. Companion Captain of the Host, you will furnish the brethren with working tools."

The Captain of the Host hands each Candidate the apron of a Most Excellent Master, which he immediately ties on. He also hands an imitation crow-

bar to one. a pickaxe to another, and a spade to the third—all of light wood—these being the working tools of a Royal Arch Mason, and explains their moral and Masonic uses as follows:

WORKING TOOLS.

C. of H.: "The working tools of a Royal Arch Mason are the Crow, Pickaxe and Spade, and may be thus explained:

"The Crow is used by operative Masons to raise things of great weight and bulk; the Pickaxe to loosen the soil and prepare it for digging; and the Spade to remove rubbish. But the Royal Arch Mason is emblematically taught to use them for more noble purposes. By them he is reminded that it is his sacred duty to lift from his mind the heavy weight of passions and prejudices which encumber his progress toward virtue; loosening the hold which long habits of sin and folly may have had upon his disposition, and removing the rubbish of vice and ignorance which prevents him from beholding that eternal foundation of truth and wisdom upon which he is to erect the spiritual and moral temple of his second life." (Royal Arch Standard, p. 82.)

C. of H. (continuing): "Brethren, follow me."

Passing out single file by way of the north-east wing curtain, the Principal Sojourner bringing up the rear, they are led to one of the outer rooms— the one in which is located the entrance to the "Secret" or "Deposit Vault," where the Captain of the Host leaves them in charge of the Principal Sojourner, and returns to his station in the Chapter.

This Secret Vault is simply what its name indicates, a vault about seven by five feet, built into the room below, as may be most convenient, and entered from above by a trap door. It is essential to the proper working of the Royal Arch degree, but because of the location of some Chapters is not always used.

The Candidates now make a pretense of digging and hauling away rubbish, under the guidance of the P. S., and soon discover the trap door leading to the secret vault. Lifting this up, they see a temporary Arch with a Keystone in the center. Removing the Keystone and consulting a few seconds as to what they ought to do, conclude to bring it up to the Grand Council. Returning by way of the Vails the

P. S. gives seven knocks at 1st Vail. (** **
** *.)

M. 1st V.: "Who comes here?"

P. S.: "Workmen from the ruins, with discoveries."

M. 1st V. (to M. 2nd V.): "Workmen from the ruins, with discoveries."

M. 2nd V. (to M. 3d V.): "Workmen from the ruins, with discoveries."

M. 3d V. (to R. A. C.): "Workmen from the ruins, with discoveries."

R. A. C. (to C. of H.): "Workmen from the ruins, with discoveries."

C. of H. (to E. H. P.): "Workmen from the ruins, with discoveries."

E. H. P.: "Let them enter."

C. of H. (to R. A. C.): "Let them enter."

R. A. C. (to M. 3d V.): "Let them enter."

M. 3d V. (to M. 2nd V.): "Let them enter."

M. 2nd V. (to M. 1st V.): "Let them enter."

M. 1st V. (to Candidates): "Enter."

The Candidates enter in single file, and preceded by the P. S., pass through the Vails and stand before the Council, one of them bearing the Keystone.

The King recites or reads from the Monitor, Zach. iv: 6-10.

King: "This is the word of the Lord unto Zerubbabel, saying, Not by might nor power, but by my Spirit. Who art thou, O great Mountain? Before Zerubbabel thou shalt become a plain; and he shall bring forth the head-stone thereof with shouting, crying grace, grace unto it. Moreover, the word

of the Lord came unto me saying, the hands of
Zerubbabel have laid the foundation of this house,
his hands shall also finish it, and thou shalt know
that the Lord of hosts has sent me unto you. For
who hath despised the day of small things? for they
shall rejoice and shall see the plummet in the hand
of Zerubbabel with those seven." (Royal Arch
Standard, p. 83.)

P. S.: "Excellent High Priest, King and Scribe,
we repaired to the place as directed, and wrought
diligently for three days without discovering anything of importance, except passing the ruins of
several columns of the different orders of architecture. On the fourth day, we came to what we supposed to be an impenetrable rock; one of my companions, on striking it with a crow, observed that
it returned a hollow sound;
whereupon we redoubled
our efforts, and on removing more of the rubbish
found it to be the top of
an arch, from the apex of
which, with much diffi-
culty, we succeeded in
raising this stone of curious form, and having en-
graved upon its side cer-
tain mysterious characters.
Deeming this an important
discovery, we have brought it up and present it to
the Council for their inspection." Hands it to the
High Priest. (See annexed figure.)

The Council consult about the stone, examining it very carefully, and then the High Priest says:—

E. H. P.: "This is an important discovery. The Council are of the opinion that this is the keystone of an arch and wrought by a Mark Master Mason; and from the situation in which it was found. it will doubtless lead to many other important discoveries. Are you willing to penetrate this Arch in search of treasure?"

P. S.: "Although` the task will be attended with difficulty, and perhaps danger, yet we are willing to make the attempt. even at the hazard of our lives. to promote so noble and glorious a work."

E. H. P.: "Then go. brethren; and may the God of your fathers be with you."

The P. S. and Candidates return to the alleged ruins and widen the aperture in the Arch by removing the imitation stones. The Royal Arch cable-tow is then fastened about the body of one of the team —the bishop for instance—and he is lowered into the vault, where he discovers three small squares, and shaking the cable to the right, as a signal, he is hauled up, the cable-tow is removed, and all proceed again in single file as before to report to the Council. Arriving at the 1st Vail, the same alarm is made, the same announcement is passed to the High Priest as before, and the same order—"Let them enter"—is returned.

·Principal Sojourner and Candidates standing again before the Council, the Scribe reads from the Monitor, or repeats Amos ix: 2:

Scribe: "In that day will I raise up the tabernacle of David that is fallen, and close up the breaches thereof, and I will raise up his ruins, and I will build it as in the days of old." (R. A. Standard, 84.)

P. S.: "Excellent High Priest, King and Scribe, we returned to the place as before, and after removing some more of the stones, to widen the aperture, we fastened a cable-tow seven times about the body of one of my companions to assist him in descending, and it was agreed that should the place prove offensive to sense or health, he should shake the cable to the right as a signal to ascend; if, on the contrary, he wished to descend still lower, he should shake the cable to the left. In this manner he de- scended, and after some search discovered three squares. * The place now becoming offensive because of the moist air which had long been confined therein, he gave the signal to ascend. With these we have returned" (holding out the squares) "and present them to the Council for their examination."

The High Priest takes the little squares, and the Council examine them with profound interest, turning them over again and again, and consulting about them very earnestly; after which the

E. H. P.: "These are indeed a most important discovery; the Council are of the opinion that they are the jewels of Past Masters, and probably those worn by our ancient Grand Masters, Solomon King

* The small figure given above being a specimen of the squares brought up.

of Israel, Hiram King of Tyre, and Hiram Abiff; and from the place where found, we have no doubt will lead to still further and more important discoveries. Are you willing to penetrate the arch again, in search of further treasures?"

P. S.: "Excellent High Priest, although it will be attended with great difficulty, and perhaps danger, yet we are willing, even at the risk of our lives, to promote so noble and glorious a work."

E. H. P.: "You will then return to the scene of your labors; and rest assured that your valuable services shall not be unrewarded."

The P. S. and team return to the ruins, as before, the cable-tow is now tied seven times about the body of the third candidate, and he is lowered into the deposit vault, as his companion was. Groping about for a second or two, he discovers a little box with projecting handles, and, giving the signal, he is hauled up again, as represented in diagram p. 262, and all proceed again to the Council, bearing the little box or imitation ark along. Arriving at the entrance to the 1st Vail the P. S. gives seven knocks as before —(*** *** *.)

M. 1st V.: "Who comes here?"

P. S.: "Workmen from the ruins, with discoveries."

M. 1st V. (to M. 2nd V.): "Workmen from the ruins, with discoveries."

M. 2nd V. (to M. 3rd V.): "Workmen from the ruins, with discoveries."

M. 3rd V. (to R. A. C.): "Workmen from the ruins, with discoveries."

R. A. C. (to C. of H.): "Workmen from the ruins, with discoveries."

C. of H. (to E. H. P.): "Workmen from the ruins, with discoveries."

E. H. P.; "Let them enter."

C. of H. (to R. A. C.): "Let them enter."

R. A. C. (to M. 3d V.): "Let them enter."

M. 3d V. (to M. 2nd V.): "Let them enter."

M. 2nd V. (to M. 1st V.): "Let them enter."

M. 1st V. (to Candidates): "Enter."

This is the ceremony always used in returning from the ruins, and is repeated here so that the reader may clearly understand it.

The P. S. and Candidates, in single file, pass through the Vails as before, and arrange themselves in front of the Council.

E. H. P.: "Brethren, what report nave you to make?"

P. S.: "Excellent High Priest, King and Scribe, we repaired to the place as before, and I descended; (speaking for Candidate.) The sun had now reached its meridian height, and shone with such reful-gent splendor into the innermost

Royal Arch Due Guard

recesses of the arch, that I was enabled to dis-cover upon a pedestal in the easternmost part thereof this curiously wrought box, overlaid with pure gold, and having on its top and sides certain mysterious characters. Availing myself of this treasure, I gave the signal, and on ascending I found my hand involuntarily placed in this position—the due guard—(see cut) to protect my eyes from the intense light and heat of the sun. We have again returned to the Council and present this box for your inspection."

They place the little box upon a pedestal or stand in front of the High Priest. The Council rises and examines it with great care—all pretense, of course—and consult about it for a minute or so.

E. H. P.: "The Council are equally ignorant as yourselves as to this box and the several mysterious characters on its top and sides. Companions King and Scribe, let us open it, and by its contents we may be able to determine its true character; peradventure it may contain matter of value to the craft and to our people. Companion Captain of the Host, open this box with care and present its contents."

King: "Excellent High Priest, this may indeed prove to be the Ark of the Covenant. If so, it is not proper that less consecrated hands than those of the High Priest himself should touch it."

E. H. P.: "Your suggestion is timely, Companion King."

The High Priest himself then opens the box in a reverential manner, and takes out a scroll of parchment, which he hands to the King, who examines and passes it to the Scribe. The Scribe also examines it and reads:—

Scribe: "In the beginning God created the heaven and the earth; and the earth was without form and void, and darkness was upon the face of the deep; and the Spirit of God moved upon the face of the waters, and God said let there be light, and there was light." Gen. 1.: 1-3.

"And it came to pass when Moses had made an end of writing the words of this law in a book until they were finished, that Moses commanded the Levites which bare the Ark of the Covenant of the Lord, saying, Take this book of the law and put it in the side of the Ark of the Covenant of the Lord your God, that it may be there for a witness against thee."

Scribe: "Is not this the book of the law?" He hands the scroll to the King, who says:—

King: "This appears to be the book of the law." The King hands the scroll to the High Priest, who says:—

E. H. P.: "The book of the law, long lost, now found—holiness to the Lord"—draws the scroll across his forehead from left to right, in imitation of the due guard. He then hands it to the King, who exclaims:

King: "The book of the law, long lost, now found—holiness to the Lord"—makes the same sign—hands it to the Scribe.

Scribe: "The book of the law, long, lost, now found—holiness to the Lord"—also makes the sign. The Scribe retains the scroll and continues to examine it. While the High Priest is still examining the contents of the little box, the Scribe recites or reads:—

Scribe: (reading) "And Moses said, This is the thing which the Lord commandeth. Fill an omer of the manna to be kept for your generations; that they may see the bread wherewith I have fed you in the wilderness when I brought you forth from the land of Egypt. And Moses said unto Aaron, Take a pot and put an omer full of manna therein and lay it up before the Lord, to be kept for your generations. As the Lord commanded Moses, so Aaron laid it up before the testimony, to be kept." (Royal A. Standard, 85.)

Toward the close of this reading, the High Priest takes out of the box the "pot of manna" which he hands to the King. The latter examines it, and hands it to the Scribe, who says:—

Scribe: "Is not this a pot of manna?"—and passes it to the King.

King: "It appears to be a pot of manna"—hands it to the High Priest.

E. H. P.: "This is the pot of manna which Moses, by divine command, laid up in the side of the Ark as a memorial of the miraculous manner in which the children of Israel were supplied with that article of food for forty years in the wilderness."

The High Priest then goes on with his search of the little box, and the Scribe reads again from the scroll:—

Scribe: "And the Lord said unto Moses, Bring Aaron's rod again before the testimony, to be kept for a token." (R. A. S., 85.)

While the Scribe reads, the High Priest takes from the box a small artificial rod, representing Aaron's rod, and hands it to the King. The latter examines it as before, and hands it to the Scribe, who says:

Scribe: "Is not this Aaron's rod?"—passes it to the King.

King: "It appears to be Aaron's rod"—hands it to the High Priest.

E. H. P.: "This is the Rod of Aaron which budded, blossomed and brought forth fruit in a day,

and which Moses by divine command laid up in the side of the Ark as a testimony of the appointment of the Levites to the priesthood."

The Scribe, still retaining the scroll and examining it further, reads again:

Scribe: "And God spake unto Moses and said unto him, I am the Lord: and I appeared unto Abraham, unto Isaac, and unto Jacob by the name of

God Almighty, but by my name Jehovah was I not known unto them." (Royal Arch Standard, p. 87.)

As the Scribe in reading unrolls the scroll, a piece of paper falls out, which he picks up, examines, and says:

Scribe: "Is not this a key?" (hands it to the King, who says:)

King: "It appears to be a key." Hands it to the High Priest.

E. H. P.: "This must be a key to the mysterious characters on the top and sides of the box."

The Council compare the key with the characters on the box, and pretend to consult very earnestly. The High Priest at last, addressing the Candidates, says:

E. H. P.: "Brethren, the Council are of the opinion that this most valuable discovery is an imitation of the sacred Ark of the Covenant, containing copies of the law, a pot of manna, Aaron's Rod, and a piece of parchment having on it apparently a key —the application of which we have not yet fully determined. The original Ark built by Moses, Aholiab and Bazaleel was burnt when the first Temple was destroyed. Of this imitation the traditions of Masonry give the only authentic account. As the first Ark was the symbol of the divine presence with and protection of the Jewish people, and pledge of the stability of their nation, so long as they obeyed the commands of God, so is this copy a symbol of God's presence with us, so long as we live conformable to the precepts contained in the Book of the Law."

The High Priest then replaces all the articles within the Ark except the key.

E. H. P. "Companion Captain of the Host, you will cause the Ark to be seated within the Sanctuary." Gives three raps and all rise to their feet and sing the Royal Arch ode, as follows:

ROYAL ARCH ODE.

"Joy, the sacred law is found,
　Now the Temple stands complete,
Gladly let us gather round
　Where the pontiff holds his seat;
Now he spreads the volume wide,
　Opening forth its leaves to-day,
And the monarch by his side
　Gazes on the bright display.

"Joy! the secret vault is found,
　Full the sunbeams fall within,
Pointing darkly under ground,
　To the treasure we would win.
They have brought it forth to light,
　And again it cheers the earth;
All its leaves are purely bright,
　Shining in their newest worth.

"This shall be the sacred mark,
　Which shall guide us to the skies,
Bearing, like a holy ark,
　All the hearts who love to rise;
This shall be the corner-stone
　Which the builders threw away,
But was found the only one
　Fitted for the arch's stay."

　　　　　—Royal Arch Standard, p. 88.

While the ode is being sung the Captain of the Host and the Royal Arch Captain seat the Ark upon the pedestal standing between the altar of incense and the East. The High Priest gives one rap and all are seated again except the three Candidates and

Principal Sojourner who have been standing in front of the Council since they returned from the ruins the last time.

The High Priest again compares the key with the characters on the Ark and then addresses the Candidates.

E. H. P.: "By this key we are enabled to read these mysterious characters. Upon the three sides are the names of our ancient grand Masters, Solomon King of Israel, Hiram King of Tyre, and Hiram Abiff; on the fourth side is the date when this deposit was made by them, being the year of the world 3000. The characters on the top of the Ark form

the name of Deity in the Syriac, Chaldea and Egyptian languages, and when given as one word they form the Grand Omnific Royal Arch word.* And now as a reward for your valuable labors we will invest you with these august secrets and appoint you to be Masters of the Vails. In your obligation you swore that you would not communicate the Grand Omnific Royal Arch word, or the great and sacred name, in any other manner than that in which they might be communicated to you. The Council with the assistance of the Captain of the Host, Principal Sojourner and Royal Arch Captain will now instruct you."

* This alleged Key is never used in the Royal Arch degree, except as mentioned here. As a Cypher however it can be readily employed by unscrupulous Masonic leaders hiding behind the Royal Arch oath, to promote their own selfish ends, as was done by Aaron Burr in 1805.

The groups are then formed as at **opening** the Chapter, the High Priest, Captain of the Host and a Candidate making one group, the King, Principal Sojourner and a Candidate comprising another, and the Scribe, Royal Arch Captain and a Candidate another group. For the accurate formation of the groups read again carefully page 202. When all are ready the Excellent High Priest says:

E. H. P. "As we three did agree, in peace, love and unity, the sacred word to search, as we three did agree, in peace, love and unity, the sacred word to keep, so we three do now agree, in peace, love and unity, to raise a Royal Arch."

While repeating these lines they all balance three times three, as at opening the Chapter, then raise their right arms, thus forming a "living arch," their right feet forming a triangle, their left wrists also clasped, and the Grand Omnific Royal Arch word is pronounced by each member of a group from right to left, by syllables, and in low breath, as follows:

1st Comp.	*2nd Comp.*	*3d Comp.*
Jah...............	Bel...............	On.
...............Jah	...............Bel
On..............................	...............Jah
Bel...............	On.	

Thus each member of a group repeats the whole word, and all the syllables are repeated three times three.

E. H. P. (to groups): "Retain the same position cf your hands and feet, kneel upon the left knee"—they all kneel—"we will now communicate to you the ineffable name of Deity, the long lost Master's word." (See p. 273.)

This word is given like the preceding one, as follows:

1st Comp.	*2nd Comp.*	*3d Comp.*
Je...............	Ho...............	Vah.
...............Je	...............Ho
Vah.............................	...............Je
Ho...............	Vah.	

E. H. P.: "Arise."

All rise to their feet, groups break up, the Council return to their stations, and the officers and companions are seated by one rap of the gavel, while the Principal Sojourner and Candidates stand facing the East.

E. H. P. "Brethren, if you will now approach the East I will explain to you the signs of a Royal Arch Mason. There is no grip in this degree."

E. H. P. (continuing): "This is the due-guard of a Royal Arch Mason"—places his open right hand to his forehead palm downward, as if shading his eyes—"and alludes to the position in which you found your hand involuntarily placed while ascending from the Secret Vault. It also alludes to a portion of the penalty of your obligation" (See p. 263.)

"This is the penal sign"—draws the open hand edgewise across the forehead from left to right—"and alludes to the penalty of your obligation wherein you have sworn that you would suffer your skull to be smote off, etc. These signs are always given together as a salute to the High Priest upon entering or retiring from a Chapter of Royal Arch Masons, or when saluting a superior officer." (See figure p. 263.)

"This is the Grand Hailing Sign or sign of distress of a Royal Arch Mason"—places the backs of both hands with fingers interlocked on the top of the skull towards the forehead—"and alludes to the additional portion of the penalty of your obligation, namely, that of having your brains exposed to the scorching rays of the Meridian Sun" (See figure p. 197.)

"This is the sign of the Master of the First Vail"—gently casts the serpentine rod on the floor and picks it up again by the tail end—"and is in allusion to the sign given by the Lord unto Moses when He commanded him to cast his rod upon the ground and it became a serpent, etc." (p. 246.)

"This is the sign of the Master of the Second Vail"—thrusts his right hand into his bosom and takes it out, thrusts it in again and takes it out—"and alludes to the sign given by the Lord unto Moses when he commanded him to thrust his hand into his bosom, etc." (p. 247.)

"This is the sign of the Master of the Third Vail"—pours a little water from a glass on the floor—"and is in imitation of the sign given by the Lord unto Moses when He commanded him to take of the

water of the river and pour it upon the dry land."
(p. 249.)

"You will now be re-conducted to the place
from whence you came and be there invested with
what you had been divested of and await my further
will and pleasure."

The Candidates and Principal Sojourner stand
at the west side of the altar of incense, salute the
High Priest with the due guard and sign of a Royal
Arch Mason and retire to the Preparation room.
where the Candidates resume their coats, etc.. and
when ready return to the Chapter.

CHAPTER XI.

When the Candidates are conducted to the ante-room the Chapter usually takes a recess, during which the Tabernacles are taken down, the "rough road" and "old bridge," etc., removed, and the gavel sounding in the East at the discretion of the Excellent High Priest, all the companions and officers are again seated.

E. H. P.: "Companion Captain of the Host, you will provide our newly exalted companions with seats in front of the Council."

This order is promptly obeyed, and the Excellent High Priest delivers as much of the historical lecture of the degree to the Candidates, as he sees fit, or as time will permit. The following however is the usual

HISTORICAL LECTURE.

"Up to this time you have been addressed and have addressed each other by the title of brother or brethren. You will now be called Companions. And, Companions, I trust that it has not been an idle or vain curiosity, that merely grasps at novelty, which has induced you to be exalted to this most sublime degree of Masonry, infinitely more important than all which have preceded it. It is calculated to im-

press upon our minds a firm belief in the existence *
and attributes of a Supreme Being, and teaches a
due reverence for His great and sacred name. It
also brings to light many valuable treasures belonging to the Craft, after they had lain buried in darkness for the space of four hundred and seventy years,
and without a knowledge of which the Masonic
character is incomplete.

"You will remember that our Grand Master
Hiram Abiff lost his life for refusing to give the

Master's word. The reason why it was not in his
power to do so is now obvious, as it could not be
communicated except in a particular manner. Masonic tradition informs us there was built beneath
the Sanctum Sanctorum a Secret Vault, in which was
deposited a copy of the Ark of the Covenant, containing imitations of the pot of manna and Aaron's
rod, and also the Book of the Law. In this Secret

* Of what use is it to believe in the existence of a Supreme Being if
they don't believe what He says? The devils believe that much and more.

Vault the ark was placed on the Masonic Stone of Foundation, which, Masonic tradition says. was a perfect cube of white oriental porphyry, and on which was inscribed, in precious stones, the characters comprising the ineffable name. These characters were placed within an equilateral triangle, and all within a circle.

"After a series of important events, of which you will find a particular account in the history of the kings of Judah and Israel, for the space of four hundred and sixteen years from the consecration of the first temple to its destruction by Nebuchadnezzar, we find that in the eleventh year of the reign of Zedekiah, King of Judah, Nebuzaradan, Captain of the Guard of the King of Babylon, went up, besieged and took the city of Jerusalem, seized all the holy vessels, the two famous brazen pillars, and all the treasures of the king's house, his palaces, and his princes. He then set both the temple and city on fire, overthrowing its walls, towers, and fortresses, and totally levelling and razing it until it became a thorough desolation; and the remnant of the people that escaped the sword carried he away captive to the king of Babylon, where they remained servants to him and his successors until the reign of Cyrus. King of Persia, by whom they were liberated, and some of them returned to their own native land. Among those who returned were three of our ancient brethren, who discovered and brought to light the principal secrets of this degree after they had lain buried in darkness from the death of our Grand Master Hiram Abiff until the erection of the second

Temple, and as a reward for their valuable labors were exalted to be the three Grand Masters of the Vails. Those three worthies you have this evening had the honor to represent." (R. A. Standard.)

THE WORKING TOOLS OF A R. A. M.

"In addition to the Crow, Pickaxe and Spade are the Square and Compasses, which have been presented to your view in every degree of Masonry through which you have passed.

"The Square teaches us, as Royal Arch Masons, that God has made all things square, upright and perfect. The Compasses are used by operative masons to describe circles. All the parts of the circumference of a circle are equally near to the center. The circle, therefore, is a striking emblem of the relation in which the creature stands to his Creator. For, as all the parts of the circumference of a circle are equally near to its center, so are all creatures whom God has made equally near to Him. (R. A. Standard.)

"THE EQUILATERAL OR PERFECT TRIANGLE

on which the principal secrets of this degree were found, is emblematical of the three essential attributes of Deity—namely, Omnipresence, Omniscience and Omnipotence; for as the three equal sides or angles form but one Triangle, so these three attributes constitute but one God.

"The equilateral or perfect triangle was adopted by the ancients as a symbol of the Deity—as embracing in himself the three stages of time—the Past, the Present, and the Future. Among the Hebrews a yod, or point in the center of an equilateral triangle, was one of the modes of expressing the incommunicable name of Jehovah. For this reason, the number three has always been held in high estimation by the Fraternity. We find it pervading the whole ritual. There are three degrees of ancient craft Masonry, three principal officers of a lodge, three supports, three ornaments, three greater and three lesser lights, three movable and three immovable jewels, three principal tenets, three rounds of Jacob's ladder, three working tools of a Fellow-Craft, three principal orders of architecture, three important human senses, three ancient Grand Masters. The secrets of a Master Mason, or Master's word, were three times demanded, and the body of our Grand Master Hiram Abiff was buried three times.

"In short the allusion to the triangle may be found wherever we turn our steps in Freemasonry. It is held in still higher estimation by all Royal Arch Masons. There are three principal officers who compose the Council, three Masters of the Vails, three—and only three—can be exalted at the same time; there were three deposits in the Ark; our Altar is triangular, our jewels are triangular, and the grand Omnific Royal Arch word can only be communicated over a triangle." (R. A. Standard.)

E. H. P. (continuing): "My companions the great and sacred name of Deity is the ancient Master

Mason's word which was communicated by the Lord unto Moses at the Burning Bush, and was in use until just before the completion of King Solomon's Temple, but was lost at the death of our grand Master Hiram Abiff. This word is composed of four Hebrew characters, which you see enclosed within the triangle, corresponding in our language to J. H. V. H., and cannot be pronounced without the aid of other letters, which are supplied by the key words on the three sides of the triangle, that being an emblem of Deity. The Syriac, Chaldeic and Egyptian words taken as one is therefore called the Grand Omnific Royal Arch Word."

"The motto of Royal Arch Masonry is the same which you observe on the front of the High Priest's mitre: 'Holiness to the Lord.'"

E. H. P.: "I will now explain to you the Key to the mysterious characters pertaining to this degree, and which you have solemnly sworn never to reveal unlawfully, and to immediately destroy when

it has served its purpose. It consists of six straight lines intersecting each other at right angles, the figures thus formed when combined with a dot corresponding each to each with the twenty-six letters of our English alphabet, as follows:" (See p. 282.)

"This is the correct arrangement of the Royal Arch Key, as may be seen by comparing it with the characters on the sides and base of the triangle. (R.

Standard, p. 92.) On the left side are the letters

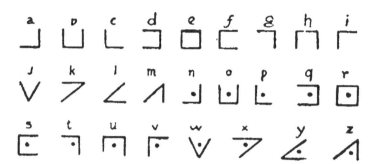

J-a-h, on the right side B-e-l, and on the base, O-n. the name of the Masonic god in Syriac, Chaldeic and Egyptian."

It is entirely optional with the High Priest whether he explains this alleged Key or not, and should he do so, can use whatever language he pleases; the explanation given above is merely a specimen of what might be said. The High Priest may also use his discretion in introducing before the following address, a description of the Breastplate, Royal Arch Banner, etc., which is purposely ömitted here, as it can be accurately found in any Chapter Monitor.

ADDRESS TO THE CANDIDATES.

E. H. P. (continuing): "And now, my companions, you have received all the instruction that pertains to our noble Craft.

"You have ascended, by regular gradations, to the summit of our sublime and royal art.

"You have been conducted around the outer courts of the Temple, viewed its beautiful proportions, its massive pillars, its starry-decked canopy, its Mosaic pavements, its lights, jewels, and furniture.

"You have been introduced into the Middle Chamber, and learned, by the example of our ancient brethren, to reverence the Sabbath-day and keep it holy.

"You have entered the unfinished Sanctum Sanctorum, and there, in the integrity and inflexible fidelity of the illustrious Tyrian, witnessed an example of firmness and fortitude never surpassed in the history of man.

"You have wrought in the quarries, and exhibited suitable specimens of your skill, and have been taught how to receive, in a proper manner, your Masonic wages.

"You have regularly passed the chair, and learned its important duties—a knowledge of which can alone qualify you to preside over the sons of light.

"You have been present, and assisted at the completion and dedication of our mystic Temple; and, for your zeal and fidelity to the Craft, have received the congratulatory title of Most Excellent Master.

"You have now witnessed the mournful desolation of Zion, the sack and destruction of the city and Temple of our God, and the utter loss, as the world supposed, of all those articles contained in the Holy of Holies.

"You have seen the chosen people of God forced by a foreign despot from the pleasant groves and peaceful vineyards of their native Israel, and dragged into captivity on the banks of the far-off Euphrates.

"But you have seen those afflicted sons of Zion visited, in the darkest night of their adversity, by a peaceful light from heaven, which guided them over rough and rugged roads to the scenes of their former glory.

"You have seen them enabled by the signet of Eternal Truth to pass the Vails that interposed between them and their fondest hopes.

"You have seen them successfully engaged in the great and glorious work of rebuilding the House of the Lord.

"And finally you have seen the sacred treasures of the first Temple brought to light, and the blessed book restored to the longing eyes of the devoted Israelites, to be the rule and guide—the comfort and support—of the people of God throughout all future time.

"And, my companions, if, in all these things, you have seen only a series of unmeaning rites—if the spirit of truth has not applied to your hearts the morals of these ceremonies—then, indeed, have we labored in vain, and you have spent your strength for naught.

"But I am persuaded to believe better things of you. I trust that you have entered into the spirit of these solemn ceremonies, and understand the full import of these interesting symbols; that all the forms and ceremonies through which you have passed, from the moment you first trod the outer courts of the Temple until your final reception within the Vails, have impressed deeply on your minds the great and fundamental principles of our time-honored institution: for then, and only then, can you justly claim the noble name of Mason; then, and only then, can you feel· that friendship, that union, that zeal, and that purity of heart which should actuate every one who would appropriate to himself the proud title of a workman that needeth not to be ashamed." (R. A. Standard.)·

E. H. P.: "If you will now stand I will deliver you the

CHARGE OF A ROYAL ARCH· MASON.

"Worthy companions, by the consent and assistance of the members of this Chapter, you are now exalted to the most sublime degree of Royal Arch Mason. The rites and mysteries developed in this degree have been handed down, through a chosen few, unchanged by time, and uncontrolled by prejudice; and we expect and trust they will be regarded by you with the same veneration, and transmitted with the same scrupulous purity to your successors.

"No one can reflect on the ceremonies of gaining admission into this place without being forcibly struck with the important lessons which they teach. Here we are necessarily led to contemplate, with gratitude and admiration, the sacred Source from whence all earthly comforts flow. Here we find additional inducements to continue steadfast and immovable in the discharge of our respective duties; and here we are bound by the most solemn ties to promote each other's welfare and correct each other's failings, by advice, admonition, and reproof.· It is a duty which we owe to our companions of this order, that the application of every candidate for admission into this Chapter should be examined with the most scrutinizing eye, so that we may always possess the satisfaction of finding none among us but such as will promote, to the utmost of their power, the great end of our institution. By paying due attention to this determination, we expect you will never recommend any candidate for our mysteries, whose abilities and knowledge of the preceding degrees you cannot freely vouch for, and whom you do not firmly and confidently believe will fully conform to the principles of our order, and fulfil the obligations of a Royal Arch Mason. While such are our members, we may expect to be united in one object, without lukewarmness, inattention, or neglect; but zeal, fidelity, and affection will be the distinguishing characteristics of our society; and that satisfaction, harmony and peace may be enjoyed at our meetings which no other society can afford."

E. H. P. (concluding): "My companions, you are now entitled to all the rights and privileges of Royal Arch Masons, and on signing your names to our By-Laws, will become members of Mohawk Chapter No. 66, in good and regular standing."

E. H. P.: "Companion Captain of the Host, you will now accommodate our newly exalted companions with seats."

ROYAL ARCH CLOSING.

The closing ceremonies are almost identical with those used in opening a Chapter, with the very important difference as to the manner of forming the groups of three, to which special attention is called elsewhere.

E. H. P.: "Companion Captain of the Host you will take the steps preliminary to closing the Chapter."

C. of H.: "Companion Royal Arch Captain when a Chapter of Royal Arch Masons is about to be closed what is your duty?"

R. A. C.: "To see that the Sentinel is at his post and the Tabernacle securely guarded.",

C. of H.: "Perform that duty and inform the Sentinel that we are about to close this Chapter of Royal Arch Masons, and direct him to guard accordingly."

The R. A. C. steps to the door, opens it, whispers to the Sentinel or Tyler that the Chapter is about to be closed, closes the door and gives seven knocks **. ** *** which are answered in the same manner by the Sentinel. Returning to his station he reports:—

R. A. C.: "Companion Captain of the Host, the Sentinel is at his post and the Tabernacle is securely guarded."

C. of H.: "Excellent High Priest your orders have been duly executed."

E. H. P.: "Companion Captain of the Host, are you a Royal Arch Mason?"

C. of H.: "I am that I am."

E. H. P.: "How shall I know you to be a Royal Arch Mason?"

C. of H.: "By three times three under a living Arch and over a triangle."

E. H. P.: "Why in that manner?"

C. of H.: "Because in that manner only can the principal secrets of this degree be communicated."

E. H. P.: "Where were you made a Royal Arch Mason?"

C. of H.: "In a legally constituted Chapter of Royal Arch Masons assembled in a place representing the Tabernacle erected by our ancient brethren near the ruins of King Solomon's Temple."

E. H. P.: "How many compose a Chapter of Royal Arch Masons?"

C. of H.: "Nine or more."

E H P.: "When composed of only nine who are they?"

C. of H.: "The Excellent High Priest and Companions King, Scribe, Captain of the Host, Principal Sojourner, Royal Arch Captain and the three Masters of the Vails."

E. H. P.: "Whom do the first three represent?"

C. of H.: "Joshua, Zerubbabel and Haggai who composed the first Grand Council assembled at Jerusalem after the destruction of the first Temple and held their meetings in the Tabernacle."

E. H. P.: "Whom do the latter three represent?"

C. of H.: "Those three of our ancient brethren who were intrumental in bringing to light the principal secrets of this degree after they had lain buried in darkness from the death of our Grand Master Hiram Abiff until the building of the second Temple, a period of 470 years, and for their valuable services were appointed Masters of the Vails."

E. H. P.: "How many Vails were there?"

C. of H.: "Four."

E. H. P.: "To what do they allude?".

C. of H.: "To the four principal tribes of Israel which bore their banners in the wilderness, Judah, Ephraim, Reuben and Dan, whose emblems were the Lion, the Ox, the Man and the Eagle."

E. H. P.: "Where were the Vails placed?"

C. of H.: "At the outer courts of the Tabernacle."

E. H. P.: "For what purpose?"

C. of H.: "To serve as a covering for the Tabernacle and stations for the guards."

E. H. P.: "Why were guards placed there?"

C. of H.: "To see that none passed but such as were duly qualified and had permission. None being admitted to the presence of the High Priest, King and Scribe sitting in Council except the true descendants of the twelve tribes of Israel."

E. H. P.: "What do the banners emblematically teach?"

C. of H.: "That when engaged in the pursuit of truth, the great object of, Masonic study, we should have the courage of the Lion, the patience of the Ox, the intelligence of the Man and the swiftness of the Eagle."

Then follows the same dialogue as in opening the chapter, about the stations of the officers, which is usually omitted, after which the High Priest says:— (See p. 194).

E. H. P.: "Companion Captain of the Host you will bring the Companions to order as Royal Arch Masons and assemble them around the altar for our devotions."

C. of H.: Three raps and all arise— "Companions you will come to order as Royal Arch Masons."— They all make the due guard as in p. 263—"You will assemble around the altar for our devotions."

The Companions assemble in a circle around the altar of incense as in opening the chapter, leaving a space at the East for the Council and Captain of the Host.

C. of H.: "Excellent High Priest your orders have been executed."

The Council and Captain of the Host fill the vacant space in the circle, all kneel on the right knee forming a chain with the right hand over the left as at opening; and the following alleged prayer is repeated by the High Priest. (See p, 199).

E. H. P. "By the wisdom of the Supreme High Priest may we be directed, by his strength may we be enabled, and by the beauty of virtue may we be incited to perform the obligations here enjoined upon us, to keep inviolable the mysteries here unfolded to us, and invariably to practice all those duties out of the Chapter which are inculcated in it. Amen."

Companions: "So mote it be."

E. H. P.: "Arise"—all rise to their feet; breaking the chain.

E. H. P.: "Companion Captain of the Host you will form the Companions in groups of three, commencing on the right."

The groups in closing the Chapter are formed as follows:—Each Companion in the group of three clasps his own left wrist with his right hand, and clasps the right wrist of the Companion at his left with his left hand—the triangle is formed with the left feet instead of the right as at opening—special attention is called to this.

C. of H. "Excellent High Priest the groups are formed."

E. H. P.: "As we three did agree, in peace, love and unity, the sacred word to search, so we three do now agree, in peace, love and unity, the sacred word to keep, until we three, or three such as we, shall with one accord raise a Royal Arch."

The G. O. R. A. W. (Grand Omnific Royal Arch word) is then repeated by each group as already explained—

1st Comp.	*2nd Comp.*	*3rd Comp.*
Jah	Bel	On.
	Jah	Bel
On		Jah
Bel	On	

The groups then balance with hands clasped as above, 3 times 3 and separate.

The Council and Captain of the Host return to their stations and the High Priest says:—

E. H. P.: "Companion Captain of the Host it is my order that this Chapter of Royal Arch Masons be now closed. This you will communicate to the Companions for their government."

C. of H.: "Companions it is the order of the Exellent High Priest that this Chapter of Royal Arch Masons be now closed. Of this you will take due notice and govern yourselves accordingly. Look to the East."

Following the lead of the High Priest, they all make the signs from Royal Arch degree to that of the Entered Apprentice inclusive. Then the raps are given, as at opening.

H. P. ***   King ***   Scribe ***
H. P. ***   King ***   Scribe ***
H. P. *   King *   Scribe *

E. H. P.: "I now declare Mohawk Chapter No. 66 erected to God and dedicated to the memory of Zerubbabel closed in due form. Companion Royal Arch Captain inform the Sentinel." Gives one rap and the Companions usually go from labor to refreshment.

CHAPTER XII.

Following is the correct Lecture of the Royal Arch degree which ought to be thoroughly committed to memory, but which very few Chapter Masons take the trouble to learn, except some of the more prominent ones, when elected to office, their apparent indifference arising from the fact that it is supposed to be learned orally. Henceforth, however, they can have no such excuse, and will doubtless avail themselves of the ready means now placed within their reach.

I shall assume that the dialogue takes place between a member of an examining committee and a Companion Royal Arch Mason desiring to visit a Chapter.

Examiner: "Are you a Royal Arch Mason?"

Visitor: "I am that I am."

Ex.: "How shall I know you to be a Royal Arch Mason?"

Vis.: "By three times three under a living arch, and over a triangle."

Ex.: "Why in that manner?"

Vis.: "Because in that manner only can the principal secrets of this degree be communicated."

Ex.: "How many compose a Chapter of Royal Arch Masons?"

Vis.: "No less than nine regular Royal Arch Masons."

Ex.: "When composed of only nine who are they?"

Vis.: "The Excellent High Priest and Companions King, Scribe, Captain of the Host, Principal Sojourner, Royal Arch Captain and three Masters of the Vails."

Ex.: "Whom do the first three represent?"

Vis.: "Joshua, Zerubbabel and Haggai who comprised the first Council assembled at Jerusalem, after the destruction of the first Temple and held their meetings in the Tabernacle erected near the ruins thereof."

Ex.: "Whom do the last three represent?"

Vis.: "Three of our ancient brethren who were instrumental in bringing to light the principal secrets of this degree after the same had laid buried in darkness from the death of our Grand Master Hiram Abiff unto the building of the second Temple, a period of 470 years, and for their valuable labors were appointed Masters of the Vails."

Ex.: "How many Vails were there?"

Vis.: "Four."

Ex.: "What was the color of the first Vail?"

Vis.: "Blue, emblematic of friendship and peculiarly characteristic of a Master Mason."

Ex: "What was the color of the second Vail?"

Vis.: "Purple, which being formed of a due admixture of blue and scarlet, is therefore placed between the first and third Vails, which are of these colors, to remind us of the intimate connection existing between Symbolic Masonry and the Royal Arch degree."

Ex.: "What was the color of the third Vail?"

Vis.: "Scarlet, and is emblematic of the fervency and zeal which should actuate all Royal Arch Masons, and is peculiarly characteristic of this degree."

Ex.: "What was the color of the fourth Vail?"

Vis.: "White, and is emblematic of that purity of life and rectitude of conduct by which alone we can expect to gain admission into the Holy of Holies above."

Ex. "To what do the four Vails taken together allude?"

Vis.: "To the four tribes that bore their banners in the wilderness, Judah, Ephraim, Reuben and Dan, whose emblems were the Lion, the Ox, the Man and the Eagle."

Ex. "Where were the Vails placed?"

Vis. "At the outer courts of the Tabernacle."

Ex.: "For what purpose?"

Vis.: "To serve as coverings for the Tabernacle and stations for the guards."

Ex.: "Why were guards placed there?"

Vis.: "To see that none passed but such as were duly qualified and had permission, none being admitted to the presence of the High Priest, King and Scribe sitting in council except the true descendants of the twelve tribes of Israel."

Ex.: "Where were you made a Royal Arch Mason?"

Vis.: "In a legally constituted Chapter of Royal Arch Masons assembled in a place representing the Tabernacle."

Ex.: "Where were you prepared?"

Vis.: "In a room adjoining the same."

Ex.: "How were you prepared?"

Vis.: "Divested of my outward apparel, hood-winked, with a cable-tow seven times about my body, accompanied by two brethren possessing like qualifications with myself, in which condition we were conducted to a door of the Chapter, and caused a regular alarm to be made by seven distinct knocks." (See p. 209).

Ex.: "To what did those knocks allude?"

Vis.: "To the seventh degree of Masonry, being that upon which we were about to enter."

Ex.: "What was said to you from within?"

Vis.: "Who comes here?"

Ex.: "Your answer?"

Vis.: "Three brethren who have been regularly initiated, passed and raised to the sublime degree of Master Mason, advanced to the honorary degree of Mark Master, have been inducted into the Oriental chair of King Solomon, and received and acknowledged as Most Excellent Masters, and now seek further promotion in Masonry by being exalted to the august degree of Royal Arch Mason."

Ex.: "What was then asked you?"

Vis.: "If it was of our own free will and accord we made the request, if we were duly and truly prepared, worthy and well qualified, if we had made suitable proficiency in the preceding degrees to entitle us to this; all of which being answered in the affirmative, we were asked by what particular right or benefit we expected to gain admission?"

Ex.: "Your answer?"

Vis.: "By the benefit of the pass."

Ex.: "Had you the pass?"

Vis.: "We had."

Ex.: "What was it?"

Vis.: "Rabboni."

Ex.: "What was then said to you?"

Vis.: "We were told to wait until the Captain of the Host could be informed of our request and his answer returned."

Ex.: "What was his answer?"

Vis.: "Let them enter and be received in due and ancient form."

Ex.: "How were you received?"

Vis.: "Under a living Arch." (See p. 214).

Ex.: "Why?"

Vis.: "To impress upon our minds in the strongest possible manner that the principal secrets of this degree can only be communicated under a living Arch."

Ex.: "How were you then disposed of?"

Vis.: "We were conducted once regularly about the outer courts of the Tabernacle, when we were met by the Captain of the Host who demanded of us who we were."

Ex.: "Your answer?"

Vis.: "The same as at the door."

Ex.: "What was then said to you?"

Vis.: "That in pursuing our intentions it would be necessary for us to travel those rough and rugged roads which all Royal Arch Masons have traveled before us, and that before proceeding further we must take a solemn obligation pertaining to this degree."

Ex.: "Have you that obligation?'

Vis.: "I have."

Ex.: "Repeat it."—(See p. 219)—obligation is usually dispensed with and instead the following question is asked.

Ex.: "What is the penalty of your obligation?"

Vis.: "Binding myself under no less a penalty than that of having my skull smote off and my brains exposed to the scorching rays of the noonday sun, etc."

Ex.: "How were you then disposed of?"

Vis.: "We were then conducted once regularly about the outer courts of the Tabernacle, when the symbol of the Burning Bush was exhibited to us." (See p. 223)

Ex.: "Why was the symbol of the burning bush exhibited to you?"

Vis.: "To impress upon our minds in the most solemn manner that the words and signs following were of divine institution and as such sacredly regarded by the children of Israel and by them transmitted to their posterity as a means by which they might be known and distinguished forever after."

Ex.: "How were you then disposed of?"

Vis.: "We were again conducted once regularly about the outer courts of the Tabernacle, when a representation of the destruction of Jerusalem took place."

Ex.: "By whom was Jerusalem destroyed?"

Vis.: "By Nebuchadnezzar, King of Babylon, who in the eleventh year of the reign of Zedekiah, King of Judah, besieged the city, captured all the holy vessels together with the two famous brazen pillars, and the residue of the people that escaped the edge of the sword carried he away captive into Babylon."

Ex. "What was the duration of their captivity?"

Vis.: "Seventy years."

Ex.: "By whom were the children of Israel liberated?"

Vis.: "By Cyrus, King of Persia, who in the firs year of his reign issued his royal proclamation, saying: Thus saith Cyrus, King of Persia, The Lord God of heaven hath given me all the kingdoms of the earth and He hath charged me to build Him an house at Jerusalem, which is in Judah. Who is there among you of all His people? his God be with him, and let him go up to Jerusalem, which is in Judah and build the house of the Lord God of Israel (He is the God) which is in Jerusalem." Ezra i: 1-3. (Royal Arch Standard, p. 71.)

Ex.: "What answer did you make to the proclamation of Cyrus?"

Vis.: "Being doubtful as to the reception we should meet from our brethren the children of Israel, we said—'But behold when we shall come unto the children of Israel, and shall say unto them the God of your fathers hath sent us unto you, and they shall say, what is His name? What shall we say unto them?'"

Ex.: "Did you pursue your journey?"

Vis.: "We did, through rough and rugged roads." (p. 235.)

Ex.: "To what do these roads allude?"

Vis.: "To the journey of the children of Israel from Babylon to Jerusalem."

Ex.: "Did you meet with any obstructions on your way?"

Vis.: "We djd, several."

Ex.: "Where did you meet your first obstruction?"

Vis.: "At the First Vail, where on making a regular demand, we heard the Master of the First Vail exclaim, who dare approach this First Vail of our sacred Tabernacle? and he, supposing the enemy to be at hand, hailed his companions, who when assembled again demanded of us who we were." (See p. 226.)

Ex.: "Your answer?"

Vis.: "Three weary sojourners who have come up to help, aid and assist in rebuilding the house of the Lord without the hope of fee or reward."

Ex.: "What was then said to you?"

Vis: "By an order of the Council now in session, issued in consequence of disturbance having arisen from the introduction of strangers among the workmen, none are permitted to engage in this noble and glorious work except the true descendants of the twelve tribes of Israel; you will therefore be careful in tracing your genealogy. Who are you?"

Ex.: "Your answer?"

Vis.: "We are of your own brethren and kin, children of the captivity, true desecndants of those noble families of Giblimites sent hither at the building of the first Temple; we have been regularly initiated, passed and raised to the sublime degree of Master Mason, advanced to the honorary degree of Mark Master, have been inducted into the Oriental chair, and received and acknowledged as Most Excellent Masters; we were also present at the destruction of the first Temple by. Nebuzaradan, by whom we were carried away captives to the King of Babylon, where we have remained subject to him and his successors until the reign of Cyrus, King of Persia, by whose recent proclamation we have been liberated and have now come up to help, aid and assist in the noble and glorious work."

Ex.: "What were you then asked?"

Vis.: "By what further right or benefit we expected to gain admission?"

Ex.: "Your answer?"

Vis.: "By the will of him who hath sent us, the God of our fathers whose name is I Am that I Am. I Am hath sent us unto you."

Ex.: "Did that gain you admission?"

Vis.: "It did, within the First Vail."

Ex.: "What did the Master of the First Vail say unto you?"

Vis.: "Three Most Excellent Masters we must have been, thus far to have come to promote this noble and glorious work, but further we could not go without his words, sign and words of explanation."

Ex.: "What were his words?"

Vis.: "Shem, Ham and Japhet."

Ex.: "What was his sign?"

Vis.: "In imitation of the sign given by the Lord unto Moses when he commanded him to cast his rod upon the ground." (See p. 246).

Ex.: "What were his words of explanation?"

Vis.: "Explanatory of that sign and are to be found in the writings of Moses, as follows:—

"And Moses answered and said, But behold they will not believe me, nor hearken unto my voice, for they will say, The Lord hath not appeared unto thee; and the Lord said unto him, What is that in thine hand, and he said, A rod; and He said, Cast it on the ground. And he cast it on the ground and it became a serpent, and Moses fled from before it; and the Lord said, Put forth thine hand and take it by the tail; and he put forth his hand and caught it, and it become a rod in his hand, that they may believe that the God of their fathers, the God of Abraham, the God of Isaac, and the God of Jacob hath appeared unto thee."

Ex.: "Where did you meet the next obstruction?"

Vis.: "At the Second Vail, where on making a regular demand we heard the Master of that Vail exclaim, Who dare approach this Second Vail of our sacred Tabernacle? Who comes here?"

Ex.: "Your answer?"

Vis.: "The same as before."

Ex.: "What were you then asked?"

Vis.: "By what particular right or benefit we expected to gain admission?"

Ex.: "Your answer?"

Vis.: "By the benefit of the words, sign and words of explanation of the Master of the First Vail."

Ex : "Did they gain you admission?"

Vis.: "They did, within the Second Vail."

Ex.: "What did the Master of the Second Vail say to you?"

Vis.: "Three Most Excellent Masters we must have been, thus far to have come to promote this noble and glorious work, but further we could not go without his words, sign and words of explanation."

Ex.: "What were his words?"

Vis.: "Moses, Aholiab and Bazaleel."

Ex.: "What was his sign?"

Vis.: "In imitation of the sign given by the Lord unto Moses when He commanded him to put his hand into his bosom." (See p. 247.)

Ex.: "What were his words of explanation?"

Vis.: "Explanatory of the sign, and are to be found in the writings of Moses, as follows:— And the Lord said furthermore unto him, Put now thine hand into thy bosom, and he put his hand into his bosom, and when he took it out, behold, his hand was leprous as snow. And He said, Put thine hand into thy bosom again, and he put his hand into his bosom again, and plucked it out of his bosom, and behold it was turned again as his other flesh. And it shall come to pass if they will not believe thee, neither hearken to the voice of the first sign, that they will believe the voice of the latter sign." (Royal Arch Standard, p. 78.)

Ex.: "Where did you meet the next obstruction?"

Vis.: "At the Third Vail, where on making a regular demand we heard the Master of the Third Vail exclaim, Who dare approach this Third Vail of our sacred Tabernacle? Who comes here?"

Ex.: "Your answer?"

Vis.: "The same as before."

Ex.: "What were you then asked?"

Vis.: "By what particular right or benefit we expected to gain admission."

Ex.: "Your answer?"

Vis.: "By the benefit of the words, sign and words of explanation of the Master of the Second Vail."

Ex.: "Did they gain you admission?"

Vis.: "They did within the Third Vail."

Ex.: "What did the Master of the Third Vail say to you?"

Vis.: "Three Most Excellent Masters we must have been, thus far to have come to promote this noble and glorious work, but further we could not go without his words, sign and words of explanation and the signet of truth."

Ex.: "What were his words?"

Vis.: "Joshua, Zerubbabel and Haggai."

Ex.: "What was his sign?"

Vis.: "In imitation of the sign given by the Lord unto Moses when He commanded him to take of the waters of the river and pour it upon the dry land." (See p. 249.)

Ex.: "What were his words of explanation?"

Vis.: "Explanatory of this sign and are to be found in the writings of Moses, as follows: 'And it shall come to pass if they will not believe also these two signs, neither hearken unto thy voice, that thou shalt take of the water of the river and pour it upon the dry land, and the water which thou takest out of the river shall become blood upon the dry land.'" (Royal Arch Standard, p. 79.)

Ex.: "What was the Signet?"

Vis.: "The Signet of Truth or Zerubbabel's Signet."

Ex.: "Where did you meet with the next obstruction?"

Vis.: "At the Fourth Vail, where on making a regular demand we heard the Royal Arch Captain exclaim, Who dare approach this Fourth Vail of our sacred Tabernacle where incense burns on our holy altar both day and night? Who comes here?"

Ex.: "Your answer?"

Vis.: "The same as before."

Ex.: "What was then asked you?"

Vis.: "By what particular right or benefit we expected to gain admission?"

Ex.: "Your answer?"

Vis.: "By the benefit of the words, sign and words of explanation of the Master of the Third Vail and the Signet of Truth."

Ex.: "What was then said to you?"

Vis.: "We were told to wait until the Captain of the Host could be informed of our request and his answer returned."

Ex.: "What was his answer?"

Vis.: "Let them enter."

Ex.: "On entering, by whom were you received?"

Vis.: "By the Captain of the Host, who conducted us into the presence of the Council, who proceeded to examine us as to our proficiency in the preceding degrees, which satisfying the Council, they asked us what part of the work we were willing to undertake."

Ex.: "Your answer?"

Vis.: "Any part, even the most servile, to promote so noble and glorious a work."

Ex.: "What was then said to you?"

Vis.: "That from the specimen of skill we had exhibited the Council were satisfied of our ability to perform any part, even the most difficult, but as it was necessary to remove some more of the rubbish from the easternmost part of the ruins in order to lay the foundation of the second Temple, we would commence our labors there, and we were cautioned to be careful to observe and preserve everything of importance, as the Council had no doubt that many valuable models of excellence lay buried there, which if brought to light would prove of essential service to the Craft."

Ex.: "What followed?"

Vis.: "We were furnished with aprons and working tools by the Captain of the Host, repaired to the place as directed and wrought diligently for three days without discovering anything of importance except passing the ruins of several columns of the different orders of architecture. On the fourth day we came to what appeared to be an impenetrable rock, but on striking it with a crow observed that it returned a hollow sound, whereupon we redoubled our efforts, and on removing more of the rubbish found it to be the top of an Arch, from the apex of which with much difficulty we succeeded in raising a stone of curious form and having engraved upon its side certain mysterious characters. Deeming that stone an important discovery we returned with it to the Council for their inspection." (See p. 258).

Ex.: "What was their opinion of the stone?"

Vis.: "That it was the keystone of an Arch and was wrought by a Mark Master Mason, and from the situation in which it was found, it would undoubtedly lead to many other important discoveries, and they asked if we were willing to penetrate the Arch in search of treasure?"

Ex.: "Your answer?"

Vis.: "Although the task would be attended with difficulty and perhaps danger yet we were willing to make the attempt even at the hazard of our lives to promote so noble and glorious a work."

Ex.: "What followed?"

Vis.: "We repaired to the place as before and after removing some more of the stones to widen the aperture we fastened a cable-tow seven times about the body of one of our companions to assist him in descending; and it was agreed should the place prove offensive to sense or health, he should shake the cable to the right as a signal to ascend, if on the other hand he wished to descend still lower he should shake the cable to the left. In this manner he descended and after some search discovered three

squares. The place now becoming offensive by
reason of the moist air which had long been confined
therein, he gave the signal to ascend, and with the
squares we returned to present them to the Council
for their examination." (See p. 262).

Ex.: "What was their opinion of the squares?"

Vis.: "That they were jewels of Past Masters
and probably those worn by our three ancient Grand
Masters, Solomon, King of Israel, Hiram, King of
Tyre and Hiram Abiff, and from the place where
found would doubtless lead to still further and
more important discoveries, and asked if we were
willing to penetrate the Arch again in search of further treasures?"

Ex.: "Your answer?"

Vis.: "Although it will be attended with great
difficulty and perhaps danger, yet we are willing
even at the risk of our lives to promote so noble and
glorious a work."

Ex.: "What followed?"

Vis.: "Being answered that our labors should
not be unrewarded, we repaired to the place as before, when I descended as before. The sun had now
reached its meridian height and shone with such
refulgent splendor into the innermost recesses of
the Arch that I was enabled to discover upon a
pedestal in the easternmost part thereof a curiously
wrought box overlaid with pure gold and having on
its top and sides certain mysterious characters.
Availing myself of this treasure I gave the signal
and on ascending I found my hands involuntarily
placed in this position (gives the due guard, p. 263)

to protect my eyes from the intense light and heat of the sun. We then returned to the Council again and presented the box for their inspection." (See p. 264.)

Ex.: "What was their opinion of the box?"

Vis.: "That it was an exact representation of the Ark of the Covenant."

Ex.: "What did it contain?"

Vis.: "A book, a pot, and a rod." (See p. 277.)

Ex.: "What was their opinion of the book?"

Vis.: "That it was the book of the law, long lost, now found, and which contained a key to the characters on the top and sides of the Ark by which we were enabled to discover that those upon three sides were the names of our ancient Grand Masters, Solomon, King of Israel, Hiram King of Tyre and Hiram Abiff. On the fourth side was the date when this deposit was made by them, being the year of the world 3,000. The characters on the top of the Ark formed the name of Deity in the Syriac, Chaldeic and Egyptian languages, and when given as one word they form the Grand Omnific Royal Arch Word."

Ex.: "What was their opinion of the pot?"

Vis.: "That it was the pot of manna which Moses by divine command laid up in the side of the Ark as a memorial of the miraculous manner in which the children of Israel were supplied with that article of food for forty years in the wilderness."

Ex.: "What was their opinion of the rod?"

Vis.: "That it was the rod of Aaron which budded, blossomed and brought forth fruit in a day and

which Moses by divine command laid up in the side
of the Ark as a testimony of the appointment of the
Levites to the priesthood."

Ex.: "How were your labors rewarded?"

Vis.: "The Council descended and imparted
to us the principal secrets of this degree, The
Grand Omnific Royal Arch Word, and the ineffable
name of Deity, the long lost Master's Word." (See
p. 271).

Ex.: "Have you any signs belonging to this
degree?"

Vis.: "I have."

Ex.: "Show me a sign."

Vis.: Gives the due-guard. (See cut p. 263)

Ex.: "What is that?"

Vis.: "The due-guard of a Royal Arch Mason."

Ex.: "To what does it allude?"

Vis.: "To the position in which I found my
hand involuntarily placed when ascending from the
Arch."

Ex.: "Show me another sign."

Vis.: Gives the penal sign—draws his right
hand edge-wise across his forehead. (See cut p. 263.)

Ex.: "What is that?"

Vis.: "The penal sign of a Royal Arch Mason."

Ex.: "To what does it allude?"

Vis.: "To the penalty of my obligation."

Ex.: "Show me another sign."

Vis.: Gives the Grand Hailing sign. (See cut
p. 197.)

Ex.: "What is that?"

Vis.: "The Grand Hailing sign or sign of distress of a Royal Arch Mason."

Ex.: "To what does it allude?"

Vis.: "To the additional penalty of my obligation."

Ex.: "What are the working tools of a Royal Arch Mason?"

Vis.: "The Crow, Pickaxe and Spade."

Ex.: "What do they morally teach?"

Vis.: "The Crow is used by operative masons to raise things of great weight and bulk; the Pickaxe to loosen the soil and prepare it for digging; and the Spade to remove rubbish. But the Royal Arch Mason is emblematically taught to use them for more noble purposes. By them he is reminded that it is his sacred duty to lift from his mind the heavy weight of passions and prejudices which encumber his progress towards virtue, loosening the hold which long habits of sin and folly have had upon his disposition, and removing the rubbish of vice and ignorance, which prevents him from beholding that eternal foundation of truth and wisdom upon which he is to erect the spiritual and moral temple of his second life." (Royal Arch Standard, p. 82.)

This ends the Lecture of the Royal Arch degree, and nothing more remains to be said except merely to remark that there are scarcely any two lodges even in the same jurisdiction, nor in fact any two Chapters that give the "secret work" exactly alike. All intelligent Masons freely admit this fact, which in the case of Chapter Masonry is all the more singular because in addition to the state Grand Chapters. there is a General Grand Royal Arch Chapter of the United States—a central body as it were—which exercises jurisdiction over all the other Grand Chapters, dictates the correct expression of the ritual, and the proper manner of conferring each degree.

In Blue Lodge Masonry, however, that is not the case, but every Grand Lodge is sovereign and independent in itself, and is the supreme head within its own jurisdiction, the traditions of Masonry and the "Ancient Landmarks" alone being the standard of government for the entire Masonic body. Blue Lodge and Chapter Masons therefore would do well to remember that none of them "know it all," the latter especially being cautioned not to conclude too hastily that this or that part of the preceding exposition is not correct, simply because they did not happen to see it so, or because of a slight verbal difference here and there in the ritual.